C000259076

Rabbiting Terriers –
Their Work and Training

Rabbiting Terriers –
Their Work and Training

David Bezzant
with John Bezzant

The Crowood Press

First published in 2006 by
The Crowood Press Ltd
Ramsbury, Marlborough
Wiltshire SN8 2HR

www.crowood.com

British Library Cataloguing-in-Publication Data
A catalogue record for this book is available from the British Library.

ISBN 1 86126 882 3
EAN 978 1 86126 882 2

Disclaimer
Every reasonable effort has been made to trace holders of copyright in illustrations reproduced in this book. Anyone holding copyright in a featured work that has not been credited is invited to contact the author, via Crowood, and we will insert a suitable credit in future printings of this book.

Typeset by Carreg Limited, Ross-on-Wye, Herefordshire

Printed and bound in Great Britain by The Cromwell Press, Trowbridge

Contents

CHAPTER ONE

The History of Rabbiting Terriers

Terriers as a group, and in particular Jack Russell types like this one, have been favourites with the British public for a long time, both as companion and pest controller.

THE POPULARITY OF TERRIERS

It has been suggested that Britain would not be Britain without its terriers, and some people have even argued that every Englishman should have a terrier. Britain is the undisputed home of terriers, and these sentiments epitomize the high standing with which the terrier group has come to be regarded in this country. It is as much at home walking along a busy city street as it is searching out a hedgerow, and has found a permanent place in the hearts of the British people.

Terriers have been friends to pauper and aristocrat alike; even royalty has fallen for their charms, perhaps the best known association being that between King Edward VII and his Fox Terrier Caesar. Undeniably brave, the terrier group has been awarded more medals for gallantry than any other breed. One of the most notable was Dick, who remained staunchly by the side of his master Surgeon Reynolds throughout the terrifying Zulu attack on Rorke's Drift. However, the terrier is most renowned and prized for its ability as a no-nonsense working dog, which is why every hunt and most countrymen possess one type of terrier or another.

A great deal has been written about working terriers, and there are still plenty of interesting books available on the subject today. The enthusiast may justifiably wonder if

The working terrier most commonly recognized is the hunt terrier. Here, the Master of the East Down is crossing Strangford Lough on his way to the kennels in 1959. The lady in the boat is his mother.

there would be room on the sagging shelves of his bookcase for yet another book on terriers, which must surely run the risk of merely copying what has been written in previous works. Even so, there is still a lack of detailed information specifically concerning the use of terriers to hunt rabbits. I can vividly remember that when my brother was given his first terrier many years ago, he searched magazines and bookshops for literature that would teach him how to train his terrier to hunt rabbits. He found books on working to rat, and books on working to fox, and even books on working to mink – but not a single word of wisdom on the subject of working terriers to rabbit.

Years later with the arrival of my first Jack Russell, I raided the bookshelves of sporting libraries to see if anything had changed. I found the occasional chapter here and there, but these were totally eclipsed by descriptions of hunt terriers, and minute details regarding pedigree evolution. It is as if using terriers specifically to hunt rabbits is

There is nothing a terrier enjoys more than having the opportunity to catch rabbits.

hardly worthy of mention, and it is rarely valued as a rewarding field sport in its own right. This book hopes to redress this situation by putting on paper what my brother and I have learnt from our years of experience of keeping and using terriers with the sole intention of catching rabbits. The reader will discover that rabbiting is just as exciting as other forms of terrier work, both for the dog and for the handler.

EARLY RECORDS OF HUNTING WITH TERRIERS

Watching terriers mark a warren, or track up and down a hedgerow in pursuit of rabbits, I have often wondered when these little dogs were first used to hunt rabbits, and by whom. Bearing in mind that the terrier is not a dog of great antiquity – such as, for example, the Saluki – and the fact that the rabbit was introduced into Britain in the twelfth century, we can ascertain the earliest possible date that rabbit hunting with terriers could have taken place. Dog historians are quite honest about the fact that accurate information concerning working terriers is scarce right up until the nineteenth century, when a greater volume of more precise records became available.

There is, however, a fleeting mention here and there that would suggest terriers at work, and these are worth looking at. The earliest of these dates back seven hundred years, when the Lord Prior of Durham voiced his anxiety about little dogs that wandered into the rabbit warrens: as a keeper of highly prized peacocks he had just reason for his concern, because one of his birds was promptly eaten by them. Could these have been crudely fashioned terriers? The consensus of opinion is that it is quite possible, and this episode with the peacock is a typically terrier-like transgression.

Later in 1389 there was a complaint made by the Commons against artificer labourers, servants and grooms. The basis of this complaint was that terriers and other dogs were being used by these people to hunt in warrens and coney-burrows when it was deemed that they should have been in church. This incident definitely mentions terriers in association with rabbit hunting. However, it must be borne in mind that such dogs were not the standardized terrier group that we see today, and probably had only size and coat type in common. We can therefore fairly confidently deduce that in Britain, hunting rabbits with a terrier-type dog dates back approximately seven hundred years.

A BRIEF HISTORY OF TERRIERS

Understanding the Terrier

The history of the terrier is strewn with guesswork, but the emphasis is definitely towards using these tough little dogs against foxes and badgers. This is hardly surprising, as it is what most of the original breeders had in mind when they were developing the different types of terrier. However, whether it was by intention or by accident, at the same time as they successfully developed a dog that excelled at work to fox and badger, they also helped to forge a type of dog that suited the person whose main interest was catching rabbits.

It is often said that a person's behaviour and character are best understood in the context of his upbringing and the way he has been treated. In like manner the character and attributes admired in terriers today are best understood against the background of their development. Although we cannot track with any exactitude the way the present terrier group evolved, it is nevertheless quite possible to explain the reasons for the terrier's inception, the purposes for which it was kept, and the conditions in which it was expected to survive.

The Terrier's Original Purpose

From day one the terrier was, as the Latin name infers, a 'dog of the earth'. It was not kept as a companion, nor was it bred for its

Most of the terriers were bred on the principle of being suitable to go to ground after the fox and badger.

conformation in the way of much prized greyhounds and mastiffs. Terriers comprised a motley bunch of small dogs with short, crooked legs, a long lithe body and a curly tail, and they were assigned the harshest of tasks against the fiercest of foes and in the worst of environments.

The requirement for an animal capable of going to ground against foxes and badgers dictated the basic shape and size of terriers at that time. They were also expected to be able to tackle any creature classed as 'troublesome vermin' due to the damage it caused to food supplies: the list included the ferocious polecat, the evasive weasel and the prolifically multiplying rat and rabbit. In order to confront these adversaries the terrier had to possess self-confidence, lightning-quick reactions, mobility in a confined space, a strong decisive bite and the intuitive sense to use these attributes effectively. Furthermore, terriers were not to show dismay at the hard bites and blows bestowed by these creatures, nor at the harshness of the injuries they encountered whilst working: broken jaws, torn flesh and even death were not uncommon. Thus the legendary toughness of terriers – something that remains a characteristic of the group today – developed out of necessity.

The terrier has faced some incredible risks during the course of its work. This one belonged to the Ullswater Foxhounds half a century ago, and was trapped in a fox earth for fourteen days. Typical of a hardy terrier, it made a complete recovery from its ordeal.

The Treatment of Early Terriers

One might have thought that the ability to work in a way that no other dog can, and in performing that essential task of pest control, would have won the terrier a veritable host of admirers. But you would be emphatically wrong. Five hundred years ago Dr Caius, physician to royalty and a redoubtable researcher, provided the material for the first book entirely on dogs, and it was revealed that terriers were kept in appalling conditions. They were forced to live in kennels on dirty straw, and had to feed on scraps which they daringly pinched from the much larger hounds. The poor creatures were left for the most part to look after themselves, and as a result were predisposed to mange, and bore a bad reputation: in disposition they were considered to be cantankerous, and were known to be only too willing to bite; as a companion, the terrier was thought fit only for the lowly stable boy. Without a doubt many of these early terriers met an unfortunate end – but in those that survived there evolved that truly indomitable spirit to triumph against all the odds.

Thus the terrier did not have the best of beginnings, and the fact that it managed to survive in the circumstances described bears testimony to the little dog's tenacity, courage and adaptability – characteristics much valued by terrier enthusiasts of all persuasions. And one of the many reasons for the terrier being given the chance to survive was its undisputed usefulness for catching rabbits.

THE SPORT OF COUNTRYMEN

By the nineteenth century terriers had built a reputation as outstanding working dogs, and finally began to receive the esteem they deserved. Working terriers of known lineage were suddenly coveted, and were treated more and more like prized possessions. The wide variety of different terriers was recognized, and standards were drawn up to safeguard their identity.

During the time between the World Wars a lot of people considered that the only terrier that qualified as a true working terrier was one that went to ground in pursuit of foxes and badgers. Some thought of the rabbiting terrier as a dog of inferior class, and one that failed to meet the criteria for other forms of work, and was consigned to rabbiting as something of a last resort. I have never subscribed to this view myself, and believe that the rabbiting terrier develops skills specific to its sport, skills that warrant the dog respect in its own right.

From the early nineteenth century to the present day the people who have always most used the rabbiting terrier are ferreters. Admittedly the professional rabbit catcher or warrener would have been familiar with the ways of dogs and ferrets; however, he would have relied overwhelmingly on steel traps to perform his job. Pest controllers, too, made

Rat catchers like this one in Glasgow at the end of the nineteenth century appreciated the value of a terrier. He may also have used it to catch rabbits when he ventured into the countryside.

use of terriers, but predominantly for dealing with rats.

Generations of countrymen during the twentieth century would have owned a terrier of one type or another. My great grandfather was a typical example. He owned a Norwich Terrier called Raft, and by all accounts they were an inseparable and mischievous pair: whether he was riding his bicycle, eating his dinner or supping a pint, Raft would be with him. At his local pub he used to wager people that they would not be able to pick up any bingo money that they dropped on the floor while the terrier was in the room. Many brave men tried, but the feisty little dog made sure that no one succeeded, and was duly rewarded with his share of a bottle of stout poured into an ashtray that the publican kept especially for the purpose. My great grandfather had at one time been an army boxing champion; he was a physically robust, plain-speaking northerner, and the little 25cm (10in) terrier was dog enough for him. Rabbit catching was at its height during his lifetime, both as a sport and profession. He would have expected Raft to catch rabbits using his own initiative, which he did numerous times (though this is not so surprising when you consider that the Norwich Terrier is one of the most capable of rabbiting terriers, and the rabbit populations of those days were truly massive).

Terrier ownership in my family can be traced back to my great grandfather who had a Norwich Terrier that went everywhere with him. When he died, the terrier would not leave his grave, despite numerous efforts to catch him, until finally he disappeared.

Having at one time worked within the National Health Service I have had the opportunity to meet a multitude of elderly gentlemen, and am astonished at how many of them enjoyed 'having a go' at rabbiting when they were children. It was considered a natural part of their rural education, and many of them recall either themselves, or one of their friends, having a terrier that would lead them on many outings after rabbits – and those they caught would not be looked down upon, but welcomed gladly for cooking by their mothers.

They also remember the fateful year when myxomatosis was introduced, and the hideous sights that followed. Quite understandably rural folk lost the will to catch rabbits, and a lot of terriers keen on rabbiting were deprived of this employment; however, they were kept for their usefulness as rat catchers and companions. My grandfather was a farm labourer who possessed an ability to grow anything that he planted; he kept a Lancashire Heeler. Typical of many men of his generation, he was not at all sentimental towards dogs. He never once bought a can of dog food, and my industrious grandmother was tasked with conjuring up meals for the

Ratting in the ricks was a common pursuit of the countryman.

dog from scraps remaining from their own meals. The heeler justified its position in the household by killing any rats gravitating towards the chicken sheds.

As the decades passed, the rabbits gradually began to recover from the plague of myxomatosis; those hunt supporters who had maintained an interest in field sports started to hunt again, and terriers once again proved their ability as rabbit catchers. Throughout many centuries terriers have enjoyed a close association with rabbits, and whether it is by assisting ferrets, working as part of a pack, or being an opportunistic hunter whilst meandering through the countryside, the terrier's affinity for such work has never been in question. Those who use terriers today to catch rabbits are proud of the fact that they are keeping alive a pastime with a long history, and in their view the rabbiting terrier remains a dog of true merit that continues to earn its keep.

Terriers continue to be of use to the rabbit catcher, especially when used in conjunction with ferrets. This terrier is perusing his morning's work.

CHAPTER TWO

What is a Rabbiting Terrier?

The identifying term 'rabbiting terrier' describes the function of a dog that could be either a particular breed amongst the many pure breeds of terrier, or a cross of more than one breed of terrier. For example, Jocelyn Lucas used Sealyhams to hunt rabbits in a pack, while a lot of ferreters have been known to employ Jack Russell crosses to the same end. As the term suggests, the work of such terriers is to pursue the rabbit, and they may do this in a variety of ways:

- in conjunction with ferrets and purse nets;
- as part of a pack of terriers or dogs of mixed breed, in a similar fashion to hounds;
- by driving rabbits from deep cover for a waiting gun, or for another breed of dog such as a Whippet.

We shall now look at each of these methods more closely in order to determine what the terrier is required to do in each case.

THE RABBITING TERRIER AND THE FERRETER

Most ferreters have a dog, and for good reason. The right kind of dog behaving in the correct way is an invaluable aid, and can dramatically affect the outcome of a day's sport. Amongst the dogs most commonly linked with ferreting are terriers, and the small Jack Russell types are seen and talked about most often in this context. They are, without a doubt, ideally suited to this kind of work, and will form a working alliance with the ferrets quite readily.

My brother and I were quickly convinced as to the usefulness of such a dog when during our early days in the sport we would observe an elderly ferreter return with rabbits from locations where we had never caught any; and at sites that we worked but only ever caught one or two rabbits, he always managed to have twice as many. One day we decided to ask him what his secret was, and he simply pointed to his dogs, an assortment of Patterdales and Russells.

Following this conversation it wasn't long before my brother procured a rascally looking terrier optimistically described as a Jack Russell, and this bundle of mischief was soon proving its proficiency for every aspect of rabbit catching, totally convincing us as to its worth. Like all good rabbiting terriers, it would:

- accurately mark occupied burrows;
- drive any surface rabbits below so they could be successfully ferreted;
- guard bolt holes that were too difficult to net;
- pursue any rabbits that managed to evade the purse nets;
- catch rabbits that hid away in, for example, hollow logs and wood piles.

It did not take my brother and me long to realize that a terrier needed to be a regular member of the rabbiting team.

HUNTING RABBITS WITH A TERRIER PACK

The word 'pack' immediately conjures up an image of the hounds and horsemen of an 'official' hunt. However, at one time a number of terrier packs were used throughout the country to hunt rabbits. Amongst the best known characters who put together a terrier pack and tramped off keenly in pursuit of rabbits are Sir Jocelyn Lucas and Brian Plummer. Both these men recount catching about sixty rabbits during a season of hunting their terriers; Plummer even managed to catch thirty rabbits during one year whilst he was simply exercising his pack. Jocelyn Lucas' pack of Sealyham Terriers was huge, being composed of more than twenty dogs, though nowadays the average pack size will rarely exceed six terriers.

The idea of using such a large terrier pack was to provide a sporting spectacle where the dogs were made to work as hard as possible in order to achieve success. This method is not, however, as effective as some other methods of rabbit hunting. In a similar fashion to the hound hunting a fox, the terrier must locate a rabbit by sight or scent and pursue it until it is caught. This is made easier by the weight and number of the terriers, and the combined advantage of their various abilities.

This type of sport is best pursued in an area where the rabbits are known to be prolific, and where there is a huge expanse of ground cover that the rabbits will use as temporary shelter, rather than taking sanctuary in a warren. For example, Plummer's premier site, where his terrier pack was most productive, was an old-fashioned rubbish tip (though this was nearly forty years ago, and nowadays he would be denied access to such a site for a variety of reasons, the chief one being health and safety). The difficulty in finding a suitable location for a large terrier pack to hunt could be one of the reasons why such packs are not as popular today as they were during the early 1900s. Other reasons are the challenge of trying to control a large group of terriers, the cost of keeping a pack and, if the terriers are owned by different people, the inconvenience of gathering them all together for a hunting foray. Well-known figures such as Tucker Edwardes and Jocelyn Lucas kept their own packs and were the breeders of the terriers they used; though it must be said, they came from a more aristocratic

A small pack is more adaptable and much better suited to rabbiting today than the large double-figured packs commonly seen during the late 1800s and early 1900s.

background than most working terrier enthusiasts do today.

Smaller packs of four to six terriers are much more adaptable to the challenges faced by the contemporary field sportsman, and therefore have a wider variety of uses; for example, they can be employed to surround an area of dense cover, or some man-made obstacle where rabbits are known to hide, and can be made to move in from all directions so as to effectively cut off the escape routes that the rabbits would normally use. Neither the cost of keeping four terriers, nor their handling, presents the problems that the much larger terrier packs used to.

The rabbiting terrier packs that I have come across are not all composed of one breed of terrier. Some include different varieties of

The Dachshund is a capable hunter of rabbits and, due to its scenting ability, worth combining with a terrier.

terrier, such as Jack Russells and Patterdales, while the more adventurous combine the skills of small hounds, such as the Beagle and the Dachshund, with terriers. They believe that the hounds have superior scenting powers and will exercise greater determination when following a trail. However, hounds can be very difficult to control; during the fifties Wendy Annette Riley regularly worked a large pack of Dachshunds, but decided to add some Border terriers to her pack in order to steady it.

RABBITING TERRIERS USED WITH OTHER DOGS

It is not uncommon for a person who is keen on working dogs to partner dogs of differing abilities in order to build a formidable team where the strength of one compensates for the weakness of others, and vice versa. This may mean putting dogs of different breeds together: thus we can see the logic of partnering the terrier with a larger, faster dog that possesses a talent for chasing rabbits across open fields.

In such situations a terrier may be used to work dense hedgerows, bramble and gorse where rabbits may be found above ground but will be able to successfully elude the clutches of larger dogs. The intention is that the terrier will make the rabbit bolt into open ground, where the bigger, faster dog can use its pace to pursue the rabbit. Depending upon the density of the ground cover, more than one terrier may be required.

This type of hunting goes right back to the early days of rabbiting with terriers, but has fallen out of favour in recent times, perhaps because too many canine partnerships were initiated where the dogs were not compatible. There are plenty of tales told of ferocious fights between terriers and lurchers, and the late Brian Plummer, who was a recognized expert on both lurchers and terriers, advised that these two should never be worked together. It would appear that sight hounds and terriers do not mix well. The exception to this rule is the Whippet, and I have heard of numerous examples of this breed being effectively teamed up with both rabbiting terriers and ferrets. Perhaps this could be attributed to the fact that in its early genealogy the whippet was infused with terrier blood, and is therefore akin to them. They are also considerably smaller than other longdogs, and are

This collie was not bred to be a rabbiting dog, but she can prove to be a great help to my terrier.

not capable of launching such a ferocious attack as their larger counterparts, even it they had a mind to.

Jocelyn Lucas hunted his pack of Sealyhams in conjunction with a German Shepherd dog, and did not encounter any problems. I know from my own experience that terriers are quite willing to team up with other dogs, both when ratting and rabbiting; however, in a lot of these teams the dogs of other breeds merely imitate what the terrier does, instead of performing their own distinctive role, which is why careful thought should be given as to what breeds to use with the rabbiting terrier if the intention is to truly broaden the scope of the work that the team can undertake whilst hunting.

My little Jack Russell makes most of the canine contribution to our rabbiting exploits, and we have noted time and again that were we to work him with a dog possessing blistering pace, the rabbits that he manages to bolt from the rough would probably be caught, rather than escape as they do at present. We have decided to try my brother's two collies, to see if they might be effective in this respect. They get on well with my Russell, which is essential, and as a trio they already take charge of all the ratting around our croft. Most importantly, the collies possess an impressive turn of speed, particularly the bitch who is built like a wisp. Furthermore, if you intend to team up a terrier with another type of dog, it is best to select a breed that will not resent the terrier being the boss.

RABBIT SHOOTING AND TERRIERS

I think that using a terrier to bolt a rabbit for the rough shooter is neither the best nor the most natural employment of the terrier's skills. Nor is it the most exciting of sports for the shooter, who may have to stand about for inordinate lengths of time waiting for a rabbit to bolt – and when it does he is expected to take a very precise shot. The terrier's instinct is to chase the rabbit as closely and for as long as possible, so there is often little distance between the terrier and the rabbit when a shot is taken; inevitably, no matter how much the terrier owner trusts the marksmanship of the shooter, he will feel extremely nervous for his dog.

It is true that terriers will not be gun-shy if trained properly, and some terriers have been used to flush game for waiting guns, and even retrieve what is shot. Even so, I still believe that terriers are not natural companions to the gun, and you need only look at the difference in character and disposition of terriers and gundogs to reinforce this view. Of all the places where I have been rabbiting I am sure that none of them would have been worked as effectively with terriers and a gun as they were with one experienced terrier and a team of ferrets.

THE ATTRIBUTES OF THE RABBITING TERRIER

Whichever of the methods already described is used to catch rabbits, the terrier will need to possess certain attributes or characteristics which, when combined, make him ideal for the job. Some of these may be instinctive, others the result of careful training, some may be anatomical, others will concern the innate spirit of the dog. For as long as there have been men and women who have worked terriers with enthusiasm, there have been different ideas on what makes the ideal terrier. For example, some favour short-legged breeds while others prefer the longer-legged terriers, and the difference of opinion has been known to become quite heated at times.

As already discussed in this chapter, I am firmly of the opinion that the best use of the rabbiting terrier is when it is worked alongside ferrets at the warren site. Consequently, this is the emphasis of the attributes that follow, and I have constantly kept in mind those terriers that are actually used regularly for rabbiting.

It is worth pointing out that just because a terrier is not in possession of one or two of the

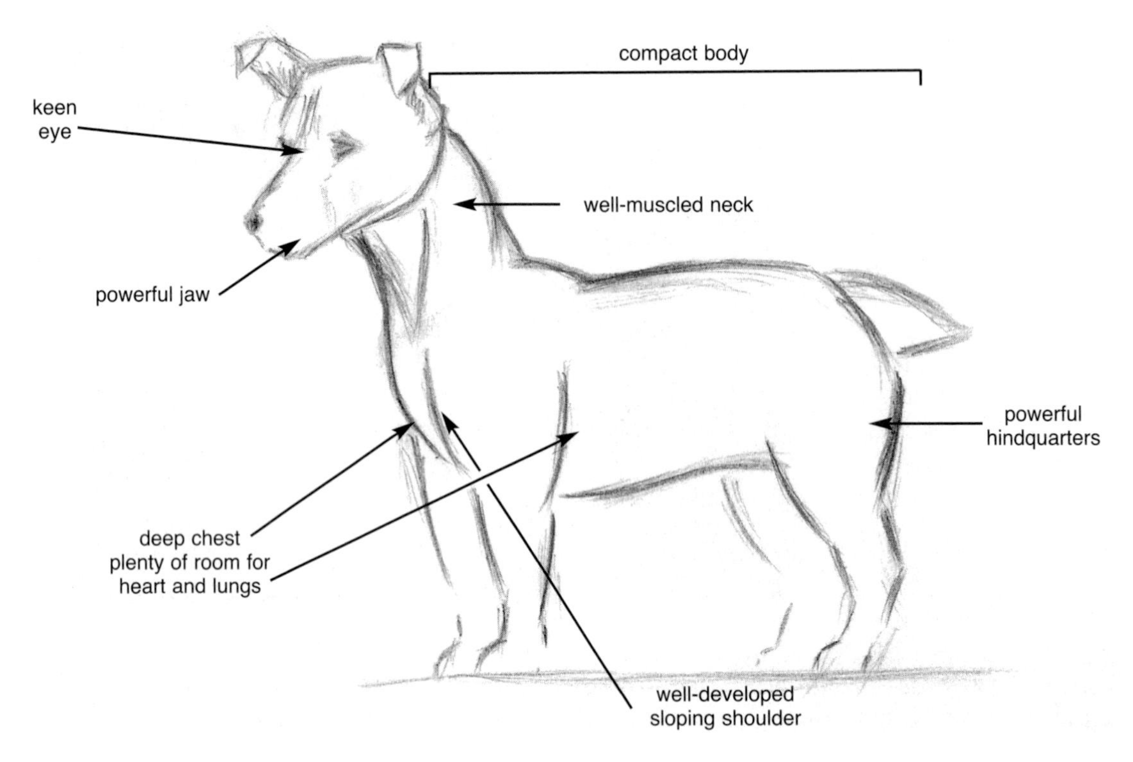

What to look for in the perfect rabbiting terrier beneath its thick protective coat. (John Bezzant)

many attributes that will be highlighted, it does not mean that it will be totally useless for rabbiting. However, if I wanted to purchase the ideal type of terrier for rabbiting, the points described below are the ones that I would be thinking about.

The Size of the Rabbiting Terrier

Within the terrier group there is a wide diversity of sizes; however, they basically fall within the bounds of the following three heights: those of 30cm (12in) or less, those of about 35 to 38cm (14 to 15in), and those of 43cm (17in) and above. Correspondingly they fall within two shapes: the smallest terrier breeds that measure about 30cm (12in) are described as 'crooked-legged' and the taller ones are described as 'straight-legged'. Some writers have even chosen to classify the terrier breeds based upon these descriptions.

Traditionally it is the crooked-legged type that earned the reputation as the best rabbiters, while the longer, straight-legged type was favoured by the hunt because they could keep pace with the hounds and men on horses. Many rabbiters of contemporary times, including my brother, continue to extol the virtues of the shorter-legged terriers, and always choose these. His simple theory is that, because the rabbit is a small animal, it will be best pursued by as small a dog as possible. Whilst he may be biased due to his terrier belonging to the short-legged brigade, there is still a degree of undeniable good sense in what he says.

Short-legged terriers definitely possess greater manoeuvrability in confined spaces, in dense cover, along hedgerows and around the warren, and are therefore able to respond more quickly to the bolting and jinking of the rabbit. On countless occasions when we have been ferreting and have been unable to gain access to place purse nets over cunningly concealed bolt holes, his terrier would be sent in

A typical example of what a rabbiting terrier should be: low on the leg, yet strong and agile, with a coat to endure the worst weather.

through the tiniest gaps to take up position in a pint-sized pocket within the hedge. Being able to maintain an alert and responsive attitude for a sustained period of time in such a confined position would not be within the scope of the bigger terriers. The terriers with short or crooked legs seem to have an instinctive knowledge of how to use their diminutive size to best effect when hunting. Taller terriers just do not seem to attempt the same antics, and for good reason.

Writing in the *Daily Mail* on 20 March 1911, a gentleman by the name of W. Bach Thomas voiced his contempt for the fashion of long, straight legs in the Fox Terrier. He considered that the Fox Terrier looked as if it were intended for Greyhound work in open country, rather than work in covert or for its

The Fox Terrier was criticized during the early 1900s because of the fashion of increasing its leg length.

original work underground down rabbit holes and fox earths. He went on to say that by making the Fox Terrier too 'beautiful', breeders were slowly destroying its ability to fulfil this original function, instead of producing the type of working terrier that was desired. The working terrier that people dreamed of was described as 'low built, strong, and full of that pluck that distinguishes the terrier class above most dogs'.

While his words may appear a little harsh, they do illustrate the fact that short legs are best for most, if not all forms of terrier work. Consequently, taking as our guide for an ideal rabbiting terrier a maximum height of 30cm (12in), we have the following candidates to choose from: the Cairn, the West Highland White, the Dandie Dinmont, the Jack Russell type, the Norfolk, Norwich, Sealyham, Skye, Yorkshire and Australian terriers. But height, or lack of it, is only one of the many characteristics of the rabbiting terrier, and it will be interesting to see how these breeds fare as we begin to look at other factors such as weight.

The Weight of a Rabbiting Terrier

Height alone cannot be considered as the ultimate indicator of a terrier's suitability for rabbit hunting; it is just one piece in the jigsaw and is not the whole picture itself. Weight is another piece of the jigsaw, and when combined with height can make the overall picture look completely different.

Take, for example, the Skye and Scottish terriers: with a height of about 25cm (10in) you would understandably think they were ideal candidates for rabbiting. However,

My brother's favourite selection of rabbiting dogs. (John Bezzant)

The modern Scottish Terrier may be considered a bit heavy for rabbiting work.

when you realize that the Scottie can weigh between 8.9 and 10.2kg (19 and 23lb), and the Skye a hefty 11.1kg (25lb), this view needs revising. What's wrong with this, you may well ask: after all, 11kg (25lb) doesn't sound that much? But when it is combined with a very small stature, it has inescapable consequences, the main one being the effect this has on the dog's overall athleticism: that is, it slows the terrier's movements and makes it less manoeuvrable. This is because, in order to accommodate the weight, the anatomical shape of the terrier must be elongated, sometimes excessively, such as 102cm ($41^{1}/_{2}$in) for the Skye Terrier.

If we compare the manoeuvrability of these long, low-slung terriers to the turning circle of cars, they are like Land Rovers, whereas

With a weight-to-height ratio of 2.5cm (1in) to 450g (1lb), the Australian is ideal for rabbiting.

Some terriers, like this Staffordshire Bull, are too substantial to be ideal rabbiting terriers.

the shorter-backed terriers are more like minis. Consequently, the longer terrier breeds suffer a distinct disadvantage when it comes to working around rabbit warrens and in dense cover, where rabbiting terriers should be at their best. Therefore we can see that while these Scottish breeds of terrier are of a desired height, their weight stops them from being the ideal general-purpose rabbiting terrier. Having said that, some of the taller terriers of more than 30cm (12in) may come back into contention because of their good weight-to-height ratio. The best recommendation is to select those with a ratio of 450g (1lb) in weight to 2.5cm (1in) in height. A good example is the Australian Terrier, with a weight of 4.5kg (10lb) and a height of 25cm (10in), giving a ratio of 2.5cm (1in) to 450g (1lb), which, as we have said, is ideal. Other breeds that measure up well are the Norfolk, West Highland White, Cairns and Jack Russell types. The taller Border and Manchester terriers also possess good statistics in this respect.

The Body Shape of the Rabbiting Terrier

Of equal importance are the anatomical dimensions that result from the terrier's combination of height and weight: its shape will have a direct influence on the speed and strength that it musters when rabbiting. We have seen already that the Skye Terrier's elongated shape hinders it as compared with other terriers. Equally, being too tall or too wide will seriously hamper the terrier's ability to penetrate deep cover where rabbits tend to lurk or seek refuge. The most obvious breed of terrier to illustrate this point is the Airedale.

I personally would not opt for a terrier over 35cm (14in) in height, and favour those that measure less – and this prompts the question of whether a terrier can be too small or too short for rabbiting. There are numerous tales told of Yorkshire Terriers that were bred so small that they could enter the warrens like ferrets and in this manner hunt the rabbit. It may sound incredible, but there are well known and reliable witnesses. Such dogs would obviously be no good if the intention was to pick up bolting rabbits in open ground, because they lack the speed and strength to bring down a rabbit quickly and cleanly. However, such postulations are largely academic due to the fact that such miniature workers are a thing of the past, and the likelihood nowadays of finding a standard Yorkshire Terrier with working blood flowing through its veins is all but impossible.

This dog shows the desired length of back for a rabbiting terrier.

Just as I would not recommend a terrier of over 35cm (14in), I would also avoid those that have an excessive body length. I have seen for myself the serious injury that can quite easily occur in long-backed dogs, and the pain that this causes. This has led to my holding the view that such a length of back is an anatomical weakness, with the risk of injury an ever-present threat. I also feel that this type of build encumbers the breed type, and is not the best body shape for either speed or strength.

I prefer a terrier with a short, strong back such as in the Cairn and Norfolk terriers. Both of these dogs can perform the typical terrier trick of sitting up on the back legs for prolonged periods of time when given the command. This shows a substantial strength of back, and I like to keep to those breeds of terrier that can perform this trick with ease.

There are those that favour a long back on short legs because of the substance it provides the terrier, and in their defence, breeds such as the Dandie Dinmont and the Skye Terrier have always had a long back, even during the time when they had an enviable working reputation. However, such a breed is rarely found in regular employment today, and there must be a reason for this; I wonder if these terriers would have been improved

Shows the typical back strength associated with most breeds of terrier. (J.H. Bezzant)

This terrier shows that, although he has only a narrow chest, he is well muscled and strong.

had they been bred with a moderate length of back.

In conjunction with a short back, I like the working terrier to possess a narrow, deep chest as opposed to a short broad one, because the former will enable it to gain access to many more confined spaces than the dog with a broader chest. A hundred years ago John Hill highlighted this fact when he wrote about a Fox Terrier called Don Cesario. According to him, the dog had such a narrow front that when he stretched himself he could go to ground in a remarkably small place. He goes on to state that short, thick, butty terriers were handicapped by a wider front and thicker shoulders, and despite being shorter on the leg, could not attempt to follow Don Cesario underground. A good guide for the breadth of chest required is to use your hand with fingers pointing towards the floor. The dog's chest should be more or less equivalent to the breadth of the palm when the dog is standing up.

It goes without saying that the rabbiting terrier should be well muscled, but this is as much dependent on food and conditioning as heredity. The areas of a well conditioned dog that will be obvious to the naked eye due to muscular development and definition are the thighs and back legs, the loins of the back, the shoulders, the chest and the neck. These muscles are easier to observe in terriers with a smooth coat, and you may have to run your hand over the body of the hairier terriers, much like a trainer does with a racehorse, in order to ascertain how well muscled it is.

The Grip of the Rabbiting Terrier

One of the requirements of the rabbiting terrier is that it possesses an effective 'grip', defined as the ability to maintain a hold for a prolonged period of time. What is desired is not necessarily a super-strong grip, but an enduring one, and in order for such a grip to develop a dog must have certain anatomical features, namely a correctly aligned jaw, being neither under- nor overshot, with strong straight teeth and a lean muscular neck. Fortunately terriers as a group have largely escaped the inbred faults that trouble other dogs, and this is equally true of the jaw. Terrier's jaws are described as either deep and powerful, long and tapering, or strong and muscular, and they are of punishing strength.

Terriers all seem to possess in common an enduring jaw strength, examples of which I

Profile of different terriers' jaws. Varied in appearance, they all possess strength and speed of movement. (D. Bezzant)

have observed time and again. For example, my Jack Russell often amuses himself with a length of rope, as only terriers can, and in so doing invariably attracts the attention of my brother's rather hefty collie dog, Pilot, who cannot resist trying to wrestle the rope free from my dog's mouth. With his far greater weight advantage he can run off dragging the terrier about like a rag doll, but sooner or later he always has to stop to readjust his grip. It may take him only a second, but the terrier will have a grip on the rope like a vice, and is off before the bemused collie knows what is happening.

This is just one of the many instances when I have noticed that the terrier can keep his grip far better than the collie. I have also observed that while the collie is obviously superior in strength due to his size and weight, he doesn't use his complete body strength to support his grip anywhere near as effectively as the Jack Russell. Perhaps it is because of his light weight or his athleticism, or simply out of sheer need because he is so much shorter, but the terrier uses every ounce of his energy in every part of his body to support his grip.

You may think all this sounds very interesting, but are beginning to wonder how it translates into useful practice when using a terrier to hunt rabbits – or, put more bluntly, why does a rabbiting terrier require an effective grip? The answer is simply that a terrier will not be able to secure sufficient hold of a rabbit if it has a weak grip. My brother's late Jack Russell bitch used to retrieve rabbits from the deepest recesses of hollowed-out logs, and was able to seize those that refused to bolt as they turned round in the mouth of a bolt hole to go back down the warren. It was only her strength of grip that enabled her to do this kind of work, and to do it quickly so as not to cause unnecessary suffering to the caught rabbit.

This does highlight another facet of the terrier's work relating to grip, and this is the speed with which it can use its jaws. Once again the terrier's small size and compact build enables him to move his head and jaw with a speed and accuracy that surpasses other breeds, and is one of the reasons why the terrier is so proficient at catching rats. The same speed and efficiency is equally desirable in the rabbiting terrier, and that is why we want the dog to possess a lean, muscular neck capable of supporting such head movement.

No matter how good the rabbiting terrier's grip is, it amounts to nothing if the dog will not respond to the voice command of the owner to release when required. Anybody whose terrier has caught a rabbit will know how unwilling they can be to relinquish their prize, but it is essential that they learn to do so in order to prevent damage to the carcass.

Although we rely heavily on the terrier's instinct to do his work, and can leave him to get on with it without interference, it is vital that we do not lose sight of the fact that the owner must oversee the activity and be confident that he can control his dog at all times simply by voice command – and this must even extend to when the terrier is using its notorious grip. It is the only safe and proper way to conduct any form of hunting – and this leads us into our next subject, which is the trainability of terriers.

The Trainability of the Rabbiting Terrier

Most of the rabbiting terrier's work is performed instinctively, which makes me wonder why it is so important that the terrier can be trained: if it will do the work itself, of what benefit is training to it? The answer is simple, but has numerous aspects to it.

Firstly, the intention of using a terrier to hunt rabbits is that its instincts be harnessed first and foremost for our benefit, and not indulged at whim. Secondly, a dog that can be trained to sit, stay and move out to the left and to the right is a much greater asset to the hunter. Thirdly, there are very few environments in which it is totally safe to work a dog today if it cannot be taught to halt and be held, unless the owner wants to rely constantly on using a lead. Fourthly, if the terrier cannot be taught to behave with certain predictable good manners it will not be very welcome on other people's property – and such property is vital for the purposes of rabbiting.

In similar fashion, let us now look briefly at why it is important to work a 'trained' terrier that is willing to respond to its owner's command when it is working. Some of the answers overlap, such as the safety of the dog: in its eagerness to pursue the rabbit an excited terrier is unlikely to notice any potential dangers, and this is quite rightly the responsibility of the handler, who is in a much better position to be aware of any risks

A rabbiting terrier must stop and listen to its handler even if it is in the midst of doing something.

to the dog. However, such far-sightedness is only of value if the terrier is willing to listen and respond to the voice of the terrier man.

A willingness to respond to voice command is also important because by it, a level of control can be maintained around the warren. The reason for wanting such control is so that the terrier can be guided to work where the more far-sighted handler believes he will be most successful. It will also enable the owner to concentrate on activities such as entering the ferret without having to worry where the terrier is, or what it is getting up to.

Most people concede that the pursuit of rabbits probably excites a terrier more than any other form of hunting, and with such an excess of stimulating sights and smells, it is important not to underestimate either the degree of training that is required to keep control of the terrier, or how compliant it is being when it stops to listen to its owner.

The Scenting Power of the Rabbiting Terrier

Although a rabbiting terrier has keen eyesight it is not a 'sight' hound, and cannot rely exclusively on its eyes to locate rabbits: it must use its nose as well, even so far as to be able to ascertain whether a warren is inhabited or not. By so doing it saves an enormous amount of time and effort, and is indispensable to the ferreter, having a direct influence on the productivity of his day.

If you were to ask most people the question, how do dogs hunt, sooner or later the word 'scent' would be mentioned, and there can't be many who have not heard something of the remarkable scenting powers of the dog. Not many of us really understand the processes so effectively at work in the dog that enable it to detect scent and to outstrip so easily man's own sense of smell. The subject has fascinated – and continues to fascinate – both lay people and professionals alike, and has been the topic of much scientific research. The verb 'to scent' is defined as 'to discern', 'to detect' or 'to track by the sense of smell'. Scent in hunting is caused by specific scent oil left on the ground by the hunted animal, and the dog is able to detect this scent because of the abundance of sensory receptors in the nose and the large area of the brain that is devoted to olfactory function.

Scent tracks are due to odoriferous particles on the ground, which impart their smell into the air that comes into contact with

The terrier uses its nose to good effect for detecting quarry.

them. Scent is carried from place to place on currents of moving air. The chief natural factor bringing about the movement of air is heat. Air that is charged with the scent of an animal will rise and move when it is heated; the same air when cooled moves by sinking. The supply of heat is commonly derived from the earth, which in turn derives its heat from the sun.

Now that we have a basic understanding of scent and scenting, we can look at how the terrier employs this sense to good effect. It will do this by first searching out the ground in a methodical manner to see if it can detect any scent; this may involve its taking some quite broad sweeps of the ground before it detects any strong smells of rabbits. When this happens, it will work round and round in small, roughly circular movements over the area where the smell is strongest, signifying this to the handler by a distinctive wag of the tail. It will follow the scent trail for as long as possible, usually to either a warren or rabbit stop. Many ferreters who have used a terrier for a long period of time readily admit their reliance upon the dog for determining if the warren is worth netting. We have learnt by experience that when the dog is not interested, no matter how promising the warren may look, it is a complete waste of time to start working it with ferrets. Likewise when the terrier shows definite interest in the most barren-looking warrens, we should take heed of his prompting, and without fail, this always proves to our advantage. If the handler acknowledges the accuracy of the canine ability to scent, then he must demonstrate this belief by trusting the dog to guide him when he is hunting rabbits.

Most, if not all terriers possess a good nose, and will instinctively follow the scent of rats and rabbits; even some hunt terriers have infuriated their masters by fastening on to the scent of rabbit when they are meant to be pursuing the fox. Obviously the scenting power of the keenest of terriers will not be as strong or enduring as that of proper hounds; nonetheless it is perfectly adequate for the rabbit-hunting man.

The Stamina of the Rabbiting Terrier

Few activities are capable of exhausting a terrier to the same degree as hunting rabbits, and they will exert all the mental and physical energy they can muster on the hunt. For the dog, rabbiting routinely means covering long distances, one minute walking, the next sprinting at full speed, turning quickly and jinking this way and that. Most athletes will tell you that such broken-paced exertion can be more exhausting then covering the ground at a consistent speed. The demands put upon the endurance abilities of the terrier's muscles are immense. For instance, the terrier is commonly expected to work the warrens and rough ground with the same vigour and commitment at the end of the day as at the start.

Stamina was bred into a lot of terriers when they were required to keep pace with hounds on a fox hunt: due to the followers' desire for a good gallop, this could mean covering many miles in one day. This is not to infer that the shorter-legged breeds of terrier are lacking in stamina; on the contrary, despite many of the Scottish breeds not being required to follow the mounted hunt, they had to climb up and then traverse the high hills, and make their way across the undulating ground characteristic of the Scottish countryside. Dogs worked in this way developed incredible powers of endurance.

Tales abound that highlight the ability of a variety of different terriers to cover great distances. These may involve crossing countries to be reunited with an owner, as in the case of a fox terrier during World War I, or simply following horses over a distance of 110km (70 miles) in one day, as in the case of Border Terriers used to hunt in Canada. Most terriers continue to possess the stamina necessary for rabbit hunting; however, I would be inclined to avoid those breeds of terrier that are now heavier and more thickset than they were when they were first bred.

Furthermore, there is nothing to equal plenty of exercise and regular work for developing the terrier's stamina; this only goes to show that, as well as being an inherited quality, stamina is affected by the way the handler manages and employs his dogs.

A terrier must have stamina. These terriers are 600m (2,000ft) up in the Lake District, and they had to rely on their own feet to get there and then be ready for work.

The Tenacity of the Rabbiting Terrier

If you were to ask people to use one word to describe the behaviour of terriers, I am sure 'tenacious' would be chosen time and again. This is not surprising for dogs that have had to overcome the indifferent and often cruel treatment of man, have had to battle with fox and badger, and deal uncompromisingly with rats. Tenacious is defined as 'holding fast', and is a characteristic of all the terriers.

Every hunting animal requires tenacity in order to succeed, and this is equally true when using terriers to hunt rabbits. Rabbits have been hunted long enough by such a variety of animals that they have developed a number of strategies to confound the hunter; and I have ferreted them long enough to know that they can be crafty, stubborn and worthy of respect. And a working terrier needs to be just as crafty and stubborn as its prey, if not more so.

When working around warrens, where terriers are normally at their best, the last thing anyone wants is a dog that gives up when it encounters difficulties or has to exert real effort to succeed. There is something about rabbits that excites dogs of any description so that they want to catch them. Many such dogs will satisfy themselves with a half-hearted chase or even pursue the rabbit until it finds some cover, when they surrender the cause. What they are lacking is 'the tenacity to hunt'. This tenacity is displayed by the dog that will hound a rabbit right to the end, no matter how long it takes. Nor will it be deterred by the harshness of the weather, the resilience of the rabbit or the abundance of seemingly impenetrable ground cover. It is this attribute of staying true to a purpose, no matter how difficult that may be, with a determination to prevail that is common to terriers, and compels them to hunt with unrestrained vigour.

Anybody who has watched working terriers go about their business will have observed how single-minded they are, reluctant to give up on anything. When my Jack Russell is

A good rabbiting terrier has to be inquisitive and have the determination to pursue its work in all environments. This little dog always hated to stop hunting at the end of a day's rabbiting.

ratting, the only way he will stop is if he catches the rat, is physically carried off, or is given some very stern commands. He would sooner work himself to exhaustion in pursuit of either rabbit or rat than be told to give up. When we took my brother's little terrier rabbiting, she would never want the day to end and was in the habit of always finding more warrens that she obviously thought we should work before going home. Her disappointment was all too evident when we ignored her.

This attitude is exactly what is wanted in a rabbiting terrier: a dog that wants to catch rabbits more eagerly than we do, and will keep at it no matter how much energy is required, or how much discomfort is involved. This is what is meant by 'tenacity' in a sporting dog; it does not mean its being keen to fight, or having a mindless disregard for its own safety.

The Intelligence of the Rabbiting Terrier

A lot of people think that intelligence and terriers do not go hand in hand. However, the multitude of various tasks that members of the terrier group have performed, and continue to do so, would suggest otherwise. The Airedale is the only British dog to have passed the strict German Schutzhund tests, which qualifies it to work as a protection or police dog. Other terriers, such as the Kerry Blue and Soft-Coated Wheaten, were used traditionally as all-purpose farm dogs, with herding sheep and cattle amongst their skills. The function of hunting also calls for the terrier to make effective use of all its skill and guile, which is a demonstration of its intelligence at work.

I have owned a terrier long enough to know that it has powers of reasoning and problem solving. If I shut him in the stables or forbid him to do something, he will think of a way out so that he can do as he pleases. Strong-willed he may be, but stupid he is not. Students of canine psychology tell us that dogs have the facility to store pleasant memories; they believe we cannot doubt that when the presence of a rat or rabbit excites the ears, eyes and nose of an experienced dog, there will be either a faint or vivid revival of pleasurable feelings of past hunting, finding and catching.

It is this power of reasoning and memory of past hunts that I want the terrier to bring to each day's rabbiting. This is because I want the terrier to give some thought to what it is

Terriers like this one have to be alert and intelligent in order to catch rabbits.

doing, and to draw from its fund of previous experiences to help it to succeed when it encounters difficulty. Hunters who have used their terriers to catch rabbits for a number of years will readily tell you that the dog becomes increasingly crafty the more it hunts.

The intelligent terrier will act sensibly, and hunt calmly and methodically. He will not be prone to over-excitement, nor will he immerse himself up to his tail in a bolt hole, trample down carefully set purse nets, or rush around with no real purpose except that of causing chaos. The rabbiting terrier knows what his job is, and possesses that confidence and cleverness that enables him to act independently and purposefully in pursuit of the rabbit. This includes recognition of how best to assist ferrets and other dogs.

The Coat Type of the Rabbiting Terrier

Throughout their history terriers have had to work without regard for the harshness of the elements, battling on in spite of driving rain, gusty winds and in some cases even snow. I often wonder what effect such continual exposure to the weather had on the dogs' health and longevity. Working sheepdogs are out in all weathers, and I have known of some that suffer from premature ageing and in particular stiffness of the joints. Doubtless the terriers would have been susceptible to the same maladies.

There are a number of different coat types within the terrier group, and these are known as rough- or wire-haired, broken-coated and smooth-coated. The rougher the coat, the better the terrier appears to be protected from the weather.

The smooth-coated types are without doubt the most vulnerable in this respect. Terriers with the smoothest of coats like my Jack Russell lack the dense coverage over all parts of the body that terriers with a rough coat have. When compared with a Scottish terrier, you would notice a definite lack of hair on the chest, belly and legs of my Russell. Essentially the dog keeps warm by trapping a layer of air against the skin, and the greater the density of hair, the larger will be the volume of air that is trapped against the skin, with the result that the dog will be able to keep warmer.

It was not an accident of breeding that those terriers that originate from the most northerly and inhospitable parts of Britain, have the most weather-resistant coats. If you could wind the clock back two hundred years, the most likely terrier that I would have owned as a resident of the hostile north-east of Scotland would have naturally been a Scottish terrier, which is ideally suited to the harsh environment. It would be advantageous for the terrier man if the counties of Britain continued to breed these terriers that originated within their boundaries, because the standard of such dogs was based upon working in the local environment.

However, many of the counties after which terriers were named no longer boast indigenous terriers with a capacity for work. Consequently those terrier breeds that have clung to their working heritage have been distributed to environments quite different to those in which they were originally bred to work. This is of particular interest to us because rabbiting is a field sport, which can be practised when the weather is at its worst and the terrier's ability to cope with the cold and wet is put to the ultimate test.

I freely admit that I have not yet come across a terrier that is put off by the weather when it is actively pursuing a rabbit. My own Jack Russell will ignore the raging wet torrents that would normally have him running for shelter, if he has the scent or sight of a rabbit in range. For the benefit of his well-being, I have to place a limit on what he should endure as a smooth-coated terrier. This is why a lot of rabbit hunters prefer a terrier with a broken coat, believing they are capable of coping better with more severe weather. However, the owner can help his dog by adopting a number of strategies: I will discuss these later, and by adopting them myself, I have been able to protect my dog from the effects of the worst of weather conditions.

The broken coat is without doubt the best for the cold and wet weather, but it does have its drawbacks, as I saw in the case of my brother's terrier. Unlike smooth-coated terriers, they can get stuck in brambles or on barbed wire, due to their longer coat – although in fairness, this rarely stops them, and usually only halts them momentarily. My Jack Russell never gets his hair caught on anything, but I have noticed that sometimes, at the end of a day's work, he may be afflicted with a number of scratches to his chest and belly. My brother's broken-coated terrier, on the other hand, often went through some horrendous-looking stuff and never damaged

The hair of a broken-coated terrier protects it from thorns and briers, but sometimes they can get stuck for a moment.

herself, which was thanks to her dense coat. I have even seen a ferret fasten hold of the terrier without causing any injury because all it had in its grasp was the dog's hair. There are arguments for and against each coat type and, as we shall see later, diet and management have an important part to play. However, if you were aiming for the ideal rabbiting terrier you would probably be best advised to select a broken-coated terrier.

Handling the Rabbiting Terrier

During the course of day's rabbiting you will need to help your terrier negotiate all manner of obstacles, from barbed-wire fences to bubbling brooks. You should bear in mind that you will probably have a carrying box and a bag full of nets on your back, and possibly a spade or stick in one hand. In such cases the easiest way to lift the terrier over an obstacle is by the scruff – the loose skin at the back of the neck. Some people dislike scruffing, but with a rabbiting terrier it is not just desirable that the terrier is amenable to being scruffed, it is essential.

During its work a rabbiting terrier may sometimes get stuck and will appreciate a good tug on whatever part of its body may be reached: this may be a leg, a bit of scruff or even the tail – which is one reason why I like terriers to have a tail long enough for me to got hold of it, if necessary. My brother once extracted his Jack Russell from a tricky situation by her ear. Terriers that take offence and try to nip when they are handled in a sometimes unavoidably rough manner are no use for rabbiting.

The Hearing Ability of the Rabbiting Terrier

The dog's hearing is much more acute than man's, and proves effective in a number of roles that it performs. This is equally the case for the terrier when it is used for rabbiting, and especially when this is in conjunction with ferrets. The hearing ability of those terriers with ears that stand up is probably slightly superior to those with folded ears.

The experienced terrier will use this sense to ascertain what is going on below ground, and if you are waiting for a rabbit to bolt, the terrier will be able to indicate where it will exit moments before it happens – and this is all thanks to its hearing. The best terriers will also be able to detect where a ferret is held up below ground, just like a ferret

A terrier must use its ears, eyes and nose to investigate its surrounds.

This old postcard illustrates the happy and affectionate character common to well-trained terriers.

locator. The terrier seems to possess an intuitive knowledge of the ways of subterranean animals, and combines this skill with its sense of hearing when it is hunting rabbits.

My brother also counts adaptability, good manners, and being happy about car travel amongst desirable characteristics that should be demonstrated by the rabbiting terrier; however, they are attributes that may be taught by a wise and conscientious owner.

Adaptability

The prime function of the rabbiting terrier is obviously to catch rabbits, but often it must also live with you in your home with the family, which may include children. Therefore, a good rabbiting terrier must be both social and adaptable. Most terriers are devoted companions and good friends to all members of the family, and those that snap at children or visitors, and kill family pets, are a disreputable minority and should be avoided. My brother and I are the fourth generation in our family to own terriers, and they have proved reliable and trustworthy to both our forbears and ourselves. I remember getting some ferrets from a fellow my brother used to know, who also had a rather nice Jack Russell type that we were admiring. Weeks later my brother heard that this very same terrier had gobbled up the man's ferrets when his back was turned.

Some people think that this kind of behaviour is only to be expected in a working terrier; however, I believe otherwise, and from my own observations of some very capable working terriers would suggest that they are totally reliable and, more often than not, better behaved than pet terriers. A rabbiting terrier will work to the utmost of its ability in the field one day, and then fit nicely into the family routine the next. Although good behaviour depends a lot on correct training, it does no harm whatsoever to purchase a pup that comes from a line of friendly and adaptable terriers.

Correct Behaviour with Livestock

The rabbiting terrier must be dependable when in contact with livestock, whether this be cows, sheep, horses, goats or chickens. The reason for this is obvious, as most rabbiting takes place in close proximity to at least some of these animals: one wrong look from your terrier may be enough for a farmer to ask you to leave his property, and he surely will if the terrier decides to scatter sheep to the four corners of a field or have a crafty nip of a cow's hock. Once again, correct behaviour is essentially a product of training.

Travel

Very few people have good rabbiting within walking distance of their home, and it is therefore essential that the rabbiting terrier is comfortable with being transported in a motor vehicle. My brother believes that some cars are better than others for dogs to travel

My terrier is amongst the most sociable of dogs with both people and other animals. Here he is playing with my mother's spaniel.

in, and puts Land Rovers at the top of the list of most suitable vehicles because the vents under the windscreen, and sliding side windows, provide excellent airflow, which prevents the dog overheating and reduces the likelihood of it feeling sick. I have known some terriers that love going in the car and others that tolerate it. My Jack Russell really does not like going in cars, no matter how big or grand they are, and he will show his protest by hiding in the smallest recess he can find and by looking miserable. However, he is able to travel long distances, one time from South Wales to north Scotland, without being either sick or causing trouble. Being able to travel, whilst not necessarily loving it, is all that is required of the rabbiting terrier when it comes to cars.

Nowadays a rabbiting terrier, like the one standing in the back of the pick-up, must be able to travel by car.

CHAPTER THREE

The Breeds of Rabbiting Terrier

Not all terriers are ideal for hunting rabbits: some are too tall and others too wide, some are too long and others too heavy. Numerous members of the terrier group were bred without any thoughts of rabbiting being paramount, and a brief look at those breeds that would not be classed as candidates for being superlative rabbiting terriers clearly highlights this. Thus the Airedale was bred to assist in hunting the otter, while the Staffordshire and bull terriers were bred for fighting. The Soft-Coated Wheaton and Kerry Blues were bred as all-purpose farm dogs, and the Fox Terrier was bred to be a servant of the hunt and therefore has long legs more suited to following horses than going to ground.

This leaves us with the breeds that have been most closely associated with hunting rabbits during the last couple of hundred years. These are the shortest of the terriers, traditionally known as short- or crooked-legged, and include the Australian, Cairn, Dandie Dinmont, Scottish, West Highland White, Skye, Sealyham, Norfolk, Norwich, Border, Patterdale, Jack Russell type and Yorkshire

A Wire-Haired Fox Terrier. During the nineteenth century these were amongst the most popular working terriers. (John Bezzant)

terriers. Some of these breeds are included primarily because of their historic use and raw potential, rather than their contemporary exploits. Others, like the Jack Russell type in particular, have their place on the list purely on the basis of their past and existing working credentials. I have selected for discussion the Bedlington, Irish and Manchester terriers because there are recent accounts detailing the talent of these breeds for catching rabbits. I must also mention the dachshund, not because I have gone mad, but because it is generally considered to be as much terrier as hound, and for the right person would make a very capable rabbit catcher.

It may help our understanding if we redefine the rabbiting terrier as a 'warren dog' instead of an 'earth dog', as the name 'terrier' implies. By excluding some breeds of terrier from our discussion, I am in no way demeaning their usefulness, but am questioning their suitability for rabbiting. If I was in need of a dog for protection I would look for an Airedale, or if I wanted a terrier for the farm I would acquire a Kerry Blue. In like manner, in desiring a dog to work the warren, common sense dictates that my selected list will present the best candidate.

Everybody has their favourite type or breed of terrier, and can prove reluctant to show partiality to any other dog. Once a hunter has found a terrier that suits his needs or works according to his requirements, he will tend to remain loyal to that breed, sometimes for a whole lifetime.

Let us now look in more detail at the individual breeds, keeping in mind the model of the ideal rabbiting terrier outlined in the previous chapter as an indicator of the particular strengths and weaknesses of each breed of terrier.

THE JACK RUSSELL TERRIER

The Jack Russell Type

Although it is not the oldest breed of terrier, the Jack Russell is probably the best known terrier in the world and the most popular within these shores. Nearly every family in the country will have owned a Jack Russell at one time or another. Most huntsmen will have seen them at work, and every sensible countryman would gladly make room in his home and heart for a good example. Their abundance means that they are quite easy to obtain and need not cost a fortune – a

A typical example of a terrier that can easily be found throughout Britain, and is ideal for rabbiting.

point that I shall return to later in the chapter.

Before we continue, it may be as well to define the type of terrier that is being referred to when the term 'Jack Russell' is used. For a long time in this country it described a predominantly white-bodied working terrier that stood between 25 to 38cm (10 to 15in) tall. Many people believed that this embraced too wide a range of terriers, some of which were not true to the original specification of the parson after whom they were named. Consequently it was decided that the denomination be divided into two categories, those around 35cm (14in) becoming known as the 'Parson's Jack Russell terrier', and officially recognized by the Kennel Club as the truest descendants of the original strain. The remainder, measuring between 25 and 33cm (10 and 13in), are still known as 'Jack Russells', but are not allowed the prefix of 'Parson'. It is these terriers that typically encapsulate what the majority of the public think of when the name of the breed is mentioned, despite the fact that they do not enter shows and cannot be issued with certification.

It is from this second group that the largest number of rabbiting terriers has emerged, and a lot of terrier men breed their own style of Jack Russell, which is often shorter and smaller than the Parson's standard. However, they do not seem to have suffered because of this, and remain in possession of desirable working qualities. They also, surprisingly, have a longer life expectancy than the Parson's Jack Russell.

My brother maintains the view that an understanding of the man known as 'Jack Russell' will help us to understand the many terriers that share his name. With this in mind, he wrote an article from which the following is quoted:

Who was Jack Russell?

It is interesting to note that as a young man, Jack Russell was duped into buying a horse that was said to be five years old, when in fact it was only two. An inspection of the teeth would have revealed this, but Jack Russell at this time, was lacking considerably in basic equine knowledge. However, in later life he was commonly regarded as something of a horse expert, which makes me wonder what happened in the intervening years to occasion such a change. I have my own theory, based upon what is known about Jack Russell, and

A portrait of Jack Russell, known as the sporting parson, and credited with developing probably the most famous of British terriers.

derived from an area of his life that is seldom considered these days, but which was essential to the make-up of the man. I am referring to the vocation that framed his character, influenced his behaviour, and compelled him to befriend people who would usually be ignored by a man of his standing.

From the outset of his ministry as a young Church of England clergyman, Jack Russell worked among the rural poor who suffered terrible deprivation. They were malnourished, lived in squalid conditions, and suffered the torment of ill health because they could not afford doctors. Often they could neither read nor write, and they commonly held many superstitious beliefs.

As part of his parish Jack Russell also had numerous gypsy encampments on Dartmoor, and the gypsies readily accepted the strong, muscular preacher. He soon came to love and respect them, and was as happy with them as when he was in the company of royalty. The gypsies have been counted amongst the most knowledgeable people in the world regarding the subjects of dogs, horses and herbal medicine, and Jack Russell's extraordinary wealth of information on these same subjects was a direct result of his long and close association with the Dartmoor gypsies.

The truth about Jack Russell is far more impressive than the multitude of tales that are told with the intention of trying to make a legend out of the man. The Parson was a man devoted to his ministry, with a real and practical passion for the people of his parish. Hunting was one of his great loves, but it was never allowed to impede his pastoral duties, and always took second place to his godly calling. This is of interest to us because, apart from proving his admirable character, it highlights the qualities of dedication and loyalty that he prized and wanted to instil in his terriers. Even today the terriers that possess his name have a dedication to duty second to none, and loyally serve their master in exactly the same manner that the minister served his parishioners, giving 100 per cent all day, every day.

The forefathers of gypsies like these could have been responsible for teaching Jack Russell about dogs and horses.

The Parson was a big, strong man who took all that life and sport could throw at him; but underneath the rustic exterior there lurked a good, kind heart, and anybody who has ever owned a Jack Russell terrier will know that the same is true of these little dogs. The terrier bears more than the Parson's name: it shares many features of his nature as well, and you would be forgiven for thinking that a little of his soul remains with us today in the terriers he established.

The Working Qualities of the Jack Russell

Jack Russells continue to possess an affinity for rabbiting, and quickly become natural allies to the ferret. They are willing to co-operate with terriers of their own kind when working, and have an insatiable appetite for sport of any description. Even when a Jack Russell is not directly descended from working stock, the instinct and talent to hunt remain strong. I noticed this when my brother acquired his Jack Russell from a non-working breeder, and despite this, she demonstrated a proficiency at rat catching and effortless application to rabbiting that resulted in my brother and me learning more from her than we taught her. For this reason they are an ideal choice for the novice terrier man.

Generally between 2.5 and 7.5cm (1 and 3in) shorter than the Parson's terrier, they have that little bit of extra manoeuvrability in tight spaces without any sacrifice of strength or power. My Jack Russell is just about 33cm (13in) tall, and is regularly employed in an old quarry where the farmer has for years dumped rubbish that the rabbits have recycled into homes and hideaways. Under wood, in between rocks and bricks, the rabbit uses it all to evade capture. My terrier moves through this rubbish with ease and often disappears out of view, in which case he lets us know what is happening by using his voice. If he is very close to a rabbit that is unable to bolt, he will grumble to himself, whereas when he is chasing a rabbit in the open he will howl like a banshee. I rather like and encourage this because it is a natural part of his working behaviour, and it also – most importantly – accurately communicates to me what is going on.

One day when we were in the quarry he located a rabbit under a partially buried door.

Jack Russells are probably the most popular and commonly used terrier by sportsmen.

As the door was basically lying on the ground, he could not get himself properly under it, which prompted my brother and me, standing at either end of the door, to tear it from its earth fastenings. We had barely begun to do this when my terrier rushed in with punishing speed and caught the rabbit. I do not believe that any other type of dog, irrespective of how quickly it can cover open ground, will be as fast as the terrier in this sort of situation, where space is at a premium.

The Jack Russell is amongst the most athletic of terriers, and I have always admired how easy they are to condition and how well muscled they become with the minimum of exercise. My brother favours the small, sturdy Jack Russell types, believing their weight and size is commensurate with the worrying of prey, which is a common method of killing amongst terriers, whether they are after rats or rabbits. My Jack Russell, on the other hand, rarely worries an animal, opting to rely on his strength of jaw and decisive bite. Having seen him in action, I am in no doubt about his outstanding grip, and have never yet seen him lose his hold on a rabbit.

As I mentioned earlier, it is considered that the non-standardized Jack Russell will, on

Terriers at play, highlighting the speed and manoeuvrability of the young terrier, which will prove to be vital when it is a working dog. Despite looking ferocious, this rough-and-tumble is all good-natured.

average, live three years longer than its kin the Parson, besides which it has no susceptibility whatsoever to inherited conditions; in fact, having such a large gene pool as it does, this terrier has a vigour that could well be beyond equal. The rabbiting man, no matter where he lives, will not encounter any great difficulty in tracking down a good strain of such terriers, and need not go to the expense of purchasing a Parson's Jack Russell Terrier. Leave these to the purists with deep pockets because they will certainly not be better, and in my experience, probably not as good rabbiters as the smaller Russells.

My Jack Russell is the product of a cross between a terrier man's dog and a whipper-in's bitch, both belonging to the same hunt in South Wales. As I have said, he measures just about 33cm (13in) at the shoulder and has a build that is ideal for warren work. He cost me all of £70 five years ago, which, if it had been doubled, would still have been considered cheap for a Parson's Jack Russell Terrier.

There are terrier men throughout the country who breed Jack Russells based upon what they consider to be the best dog for the type of sport it will undertake and the environment in which it will perform its assigned tasks. They therefore often produce a terrier that is smaller than the Kennel Club standard, and it could be argued that they are acting more in the spirit of the revered clergyman by breeding a dog to their own specification, rather than simply purchasing the established breeds.

Cheap, readily available, in possession of a working instinct and physically built for the job, the Jack Russell sounds like a paragon of all the known terrier virtues. However, there are the odd one or two problems that a prudent person should plan to avoid or deal with (*see* drawing on opposite page). The main one concerns the temperament of this breed of terrier. My brother expresses this in the following way:

> As a general rule there are two distinct types of Jack Russell temperament. The first is the mentally deranged that bark incessantly and attack the post or postman

An ideal type of terrier for rabbiting: healthy, from working parents, and easy to acquire.

> given half a chance, and then there is the laid-back fellow who would not get excited if you set fire to his tail.

Admittedly he is describing the extremes of behaviour in order to make his point, and most terrier conduct falls some way between these two extremes. Although both temperaments are capable of remarkable feats of working brilliance, I would always select my terrier from the second group and recommend that others do likewise. In my opinion a terrier that is friendly and sociable is a much more accurate reflection of the Jack Russell's essential character than the one that is overtly aggressive and yappy.

One of the things that I have always admired in Jack Russells, particularly the dogs, is their confidence. When living in the company of other breeds, such terriers invariably assume the role of leader, and it is quite amusing to watch Labradors and muscle-bound Rottweilers be guided by a 30cm (12in) tyke. New surrounds and different people do not in any way trouble him. Typically, the Jack Russell has the heart of a lion, the mind of a king and the body of a dwarf.

Like every terrier, the Jack Russell requires careful training and, from observations of my own Jack Russell, I would never claim that this is one of the breed's favourite activities. However, with perseverance the terrier can prove himself capable of basic and advanced obedience. They are amongst the easiest dogs I know when it comes to proving reliability with livestock. I have kept rabbits, ferrets, chickens, ducks, sheep, pigs, goats and ponies, and neither my brother's nor my own Jack Russell have ever troubled any of them, despite at times having received a fair amount of provocation.

The Jack Russell is suitable for both the novice and experienced worker. He will soon prove his worth around the warren, and his eagerness and commitment are a pleasure to watch. The Jack Russell is the most frequently used terrier for rabbiting and, as we have seen, this is because of good reasons. Both in the field and at home he is a loyal companion

A Jack Russell Terrier. Probably the best all-round terrier available today. (John Bezzant)

and, in my opinion, overall the best breed of rabbiting terrier available today.

BREEDS OF RABBITING TERRIER FROM SCOTLAND

In spite of my enthusiasm for the Jack Russell terrier, it would be quite wrong of me to give the impression that it is the only terrier available today that can be used for rabbiting. There are numerous other breeds of terrier that are regularly used to hunt rabbits, and there are even some that have all the potential but have fallen out of favour regarding their use for any form of field sport activity.

A number of factors have contributed to this demise: a change in rural pastimes and consequent lack of demand, the belief that this lack of use has resulted in the loss of working instinct, and the high price of purchasing pedigree breeds of terrier. A lot of people from both the show and sporting fraternities cannot imagine seeing these dogs in any other role than that in which they currently perform. Yet I have often heard of terriers that compete at shows – some of them Kennel Club breed champions – whose favourite activity is chasing rabbits. I even know of a Sealyham Terrier that is a successful show dog, and has the habit of regularly deserting its kennel in favour of days spent in the countryside procuring its own food and surviving on its wits, much to its owner's chagrin.

Neither pedigree status nor the show ring need necessarily mean the destruction of a terrier's true worth. Of the breeds of terrier that have been overlooked regarding their potential as rabbit catchers, and their unquestionable physical appropriateness for the task, the majority herald from Scotland, and are deserving of closer inspection.

The Cairn Terrier

The Cairn is the obvious terrier to start with, because it is generally considered to be the root stock from which all the breeds of Scottish terrier have emerged. In addition, the Cairn is the epitome of what a working terrier should be, and in particular it is the archetype of a rabbiting terrier.

Although the Cairn is the smallest of the working terriers, it quickly built a reputation

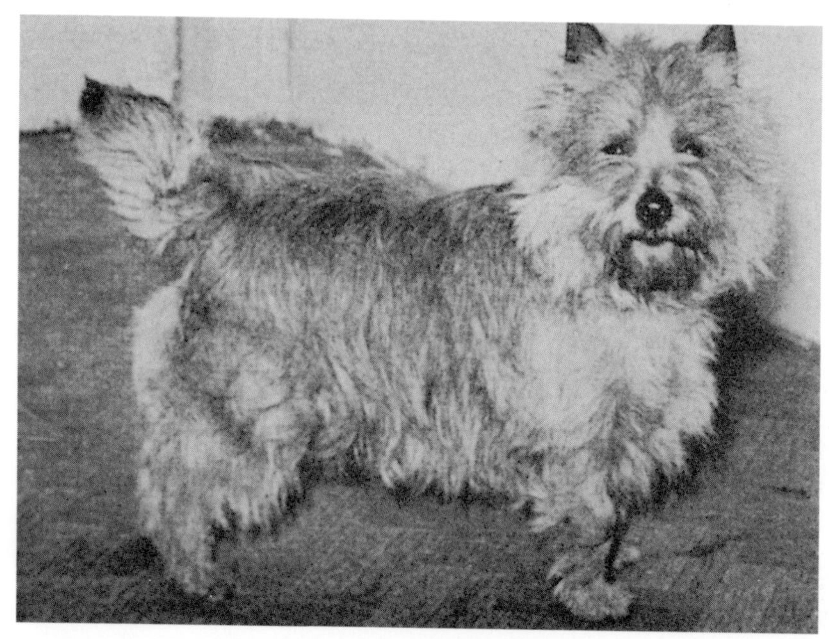

The Cairn is ideal in weight, height and coat type for rabbiting.

as one of the most fearless of dogs, which is why it is so highly esteemed amongst terrier owners from all over the world. It was bred to work the rugged terrain of the hostile Highlands, where foxes that preyed upon the flocks of sheep could easily find sanctuary. Fox hunting up there was conducted in a totally different way to the chase as it was commonly seen in England. The Highland shepherds would hire a professional fox hunter, who would work with a couple of hounds and a small pack of terriers. The hounds were used to follow the trail of the fox to its lair, the latter often located many feet under the ground beneath a rocky outcrop, where there would be an intricate honeycomb of twisting passages.

The task of the terrier was to search out this rock maze and either bolt or kill the fox. This was an extremely dangerous job, and a man armed with digging equipment was no use to the terrier that was trapped underground: unless it could find its own way out of the subterranean dungeon, it would probably never be seen again.

This is the background against which the Cairn developed: originally known as the Highland or Short-Haired Skye, it is thanks to the resolve of breeders and enthusiasts that the Cairn remains essentially the same as when first it became known. They have staunchly opposed any changes to the physical or mental make-up of the Cairn, with the result that it is just as fit for work as it was in yesteryear, and lacks only the opportunity of being widely involved in field sports today – which is undoubtedly a shame. However, it is encouraging to hear of the occasional Cairn, particularly when in Scotland, still being used in a working capacity.

The Cairn's height of 25cm (10in) makes it an ideal size for rabbiting work, and it is especially suited to penetrating the thickest of cover. The years have fortunately not seen the Cairn weighed down with additional pounds, as has been the trend with some of the other terriers, and it continues to possess all the energy necessary for a thorough day's work. The Cairn is nimble of foot and demonstrates the same manoeuvrability that it made use of when negotiating the rocky tunnels where foxes lived. Anybody who has been to Scotland and walked over the terrain where Cairns were regularly worked will be aware of how much stamina is required to simply traverse the ground, never mind the additional energy a terrier will use going up and down, backwards and forwards on a scent, and searching out every crevice in the rocks. And any visitor to Scotland in the winter will eventually pass some comment on the weather, which can be notoriously harsh and unforgiving with its combination of gusting wind, all-pervading rain and bitter cold; indeed, the Scottish weather can be of such severity that it caused Russian soldiers, who were training in the mountains, unhappily to complain that they were always cold, which they attributed to the ever-present wind chill.

Not so for the Cairn: with its hard wiry coat over a close woolly undercoat, it has proven its resistance to the worst weather imaginable – and most remarkably, the mood of the dog is not altered by the conditions. It is even reputed to be able to stay dry after a day's work in an otter burn.

Consequently, the Cairn Terrier has a lot to recommend it to the keen hunter of rabbits. In addition to its size and vigour, it has a coat that is ideally suited to winter work, and in disposition it is as game as they come. The only drawback is the lack of people breeding working Cairns, which means there is very little likelihood of being able to obtain one other than from those breeders whose market is mainly the pet and show worlds, and these animals are sold at a premium. This may not be too much of a problem if you want only one terrier, but for those who wish to use two or more, it may prove to be influential.

The Skye Terrier

The Skye Terrier is typical of a dog that has been so altered by breeding that questions arise regarding the ability of the present-day terrier to perform its original function. But make no mistake, despite appearances, the Skye was at one time every inch a working terrier, and by all accounts quite a good one.

The early Skye Terriers were described and portrayed in pictures as a dog resembling a long-bodied Cairn Terrier. As its name suggests, it had its home in the misty island of Skye, and was also commonly found throughout parts of the Hebrides and mainland Scotland. The anatomical peculiarities of a long back set upon short legs were pointed to as proof that these terriers were born to follow vermin underground. In fact it is claimed that they parallel, more than any other breed, the shape of the badger, weasel and otter, which Nature has made low in stature in order that they might inhabit earths.

It was said that if you saw a team of Skye Terriers racing up a hill after a fugitive rabbit, tirelessly burrowing after a rat, or displaying their terrier strategy around the earth of a fox, you would be forced to admit that these dogs were meant for sport, and were demons at it. It is recorded that by the year 1700, wild cats and foxes plundered the flocks in Scotland to such an extent that landlords and tenants formed associations that provided for the support of a man who would keep four foxhounds and about a dozen Skye Terriers. In this way, the wild vermin were reduced to reasonable proportions.

At one time regarded as a talented working terrier, one has to wonder what has happened to the Skye over the years. But we need look no further than the show ring, where efforts to differentiate the Skye from other breeds led to an exaggeration of its features, resulting in a broader chest, longer coat and longer body than the original. A lot of people think there is no longer still a working terrier underneath all the hair, and this is a great shame when you bear in mind how often it was used, and of how well spoken it was in centuries past.

One of the tasks of the original Skye was to catch rabbits, and it seemed quite capable of this despite the strenuous efforts required to traverse and work the difficult terrain found in western Scotland. Does this terrier have anything to recommend itself for the job of rabbiting today? Its height of 25cm (10in) is ideal, but with a body length of over 100cm (40in) it doesn't have the speed and agility of other terriers, and will be too slow for pursuing rabbits in dense cover. However, they are still considered as game dogs, and should be

A Skye Terrier. Originally this terrier was used for all forms of pest control on the steep hills of Scotland. (J.H. Bezzant)

The Skye Terrier. Its very long body affects its speed and manoeuvrability when compared with other terriers. (D. Bezzant)

able to endure all the winter weather. This is not surprising when you think that the coat of the original Skye was developed so that the dog could swim after the sea otter without becoming wet to the skin.

The Skye Terrier is an enthusiast's dog, with no one nowadays willing to vouch for any working ability in the contemporary strain. Working terrier enthusiasts are put off by the ridiculously long coat displayed by the show Skye, and the long back with its obvious potential for injury. The Skye Terrier is hardly ever seen in public, and the only reason for the acquisition by a rabbiter of such a dog would be if he wanted to try and revive a working strain of this ancient breed; but judging by the efforts necessary with some other terriers, this would prove to be a very difficult project.

The Scottish Terrier

The Scottish Terrier is a very distinguished-looking dog, and his character can be largely identified by his appearance. He has the face of a sage, the disposition of a kirk elder and the physique of the anchorman of a tug-of-war team. What is not so easy to ascertain about him, if he has only been seen within the show ring, is that underneath the over-barbered coat is the heart and desire of a working terrier.

Like the other terriers of Scotland, the Scottie, as he is known, was kept in the Highlands for his ability to confront indigenous predators, and proved himself to be a tough, hard-biting and persevering worker. His job was to bolt the foxes from the cairns, and for that he needed to be plucky, active and with plenty of voice. There were those who criticized the Scottie, claiming that his skull was over-long for the purpose of killing vermin in flight. But his admirers quickly came to his defence, pointing out that a badger's skull is long and clean, and there is no more punishing a bite than that possessed by a 'brock' (badger).

Another idiosyncratic feature of the Scottie is his unusually hard skin; this feature helped to protect him from injury when he was about his business in the hostile north-east of Scotland, and here he became extremely popular, particularly around Aberdeen, once his good points were known. The weather in this part of the country would have been as fierce as many of his opponents, but his double coat would have treated it with scorn, and is of such a texture that with a shake or two after a drenching he is nearly dry. Today we think of the Scottish Terrier as a black dog, but his coat could be coloured either steel or iron grey, brown brindle, grey brindle, black brindle, sandy or wheaten, and solid colours with no white markings were preferable.

The Scottish Terrier was not just a working terrier, he was an intimate friend of his master, the small farmer, and his wife and children. Sharing their meals and hardships, the terrier developed a very high intelligence with regard to reading his master's looks and anticipating his wishes. In fact, like his friend and companion the collie, he could do anything except talk. So peculiar is this characteristic that it is said that if a man has once owned a Scottie, he never wants another dog.

The Scottie becomes devoted to his master and wants to accompany him throughout the day and share his interests, and should these include chasing vermin, he will be particularly happy. He has a very good nose for searching out rat or rabbit; with his powerful teeth and strong paws he can dig a large hole nearly as quickly as a man with a spade, and enjoy it much more.

The Scottish Terrier has many of the qualities that one could wish for in a rabbiting terrier. Being from between 23 to 30cm (9 to 12in) tall, he is of an ideal height and has an enviable ability to stay warm whatever the weather. Probably one of the greatest attractions of the Scottish terrier is his character, and if you want a dog that gets on with its work with unnecessary fuss, a Scottie could prove to be the perfect choice.

However, I do have some concerns about the modern Scottie, chief of which is its weight, often being over 9kg (20lb). It is claimed that this heaviness is a result of the substance of bone that the Scottie possesses, but the earliest Scotties weighed approximately 7.5kg (17lb) and were of adequate substance for work. When compared with some other terriers, the Scottie does look a bit clumsy and is slower moving. He can also be rather stubborn, and is apt to think that he knows best. The Scottish terrier has not been in regular work for approaching half a century, and is rare enough today as a pet. Just as they are difficult to find, they will be expensive to buy.

When it comes to rabbiting, the Scottie still has some very strong features, such as grip, coat and intelligence; but it also has some incredibly weak ones, which are being too

The Scottish Terrier is renowned for its intelligence and has an exceptionally strong jaw.

The Scottish Terrier. Heavy boned and lacks speed and agility when compared with other breeds. (D. Bezzant)

heavy, slow and rather broad. The Scottish Terrier is very steady and so would be suitable for the novice; however, I feel that the more experienced rabbiter may resent its lack of pace. I have an interest in the breed derived from the fact that I live in a part of Scotland where, one hundred years ago, the terrier was very popular. I would very much like to see how one would fare at the sport today, and how its appearance would alter for being in working condition.

The West Highland White

The West Highland White is certainly amongst the most popular of terriers with the British public, and I would hazard a guess that today's terrier owners would be horrified if their Westies were used in the same manner as their forbears of four hundred years ago.

Their job was to go to ground and turn out foxes, otters and badgers from the strongholds in which they lived. These strongholds, located in the remote areas of the west Highlands, are more impregnable than any architect could design. The lairs lie deep in the hills under rocks of such weight that they are immovable. There is no soil or sand to dig away at with either paws or spades, and the Westie had to squeeze its body through the rocks. Negotiating this natural fortress dictated the size and shape of the terrier, because too wide of girth and it would get stuck, too heavy and it would lack the activity to jump up and down amongst the rocks where there was no room to take a run; and the terrier had to be able to use its paws to fasten on to the edges of rocks and pull itself up. An early proponent of the breed stated that 'unless a person has seen the awful masses of tumbled rocks these dogs work among, he can hardly realize the degree to which jumping and scrambling power is necessary. Without this a dog is likely to be trapped.'

The West Highland White proved itself so capable at its work that it was deemed fit as a present to be given and received from royalty. It continued its prowess right up to the early part of the last century when it was frequently used by aristocratic owners of sporting estates in Scotland, who did much to popularize the breed and publicize its hunting skills. Colonel E.P. Malcolm of Poltalloch was the most vociferous of these, and the West Highland White was for a time known as the 'Poltalloch terrier' after the strain the Colonel had developed and used on his estate.

The physical components essential to its earliest work adequately equip the terrier for

Two of Colonel Malcolm of Poltalloch's terriers. Although the picture quality is not ideal, you can still see that these dogs have the look of true working terriers.

hunting the rabbit today, and a comparison of the modern West Highland White with those of a hundred years ago shows that they have not changed drastically. Colonel Malcolm had collected white terriers that were not favoured in Scotland during the late eighteen-hundreds, and bred his strain of working terriers from them with an indifference to their ancestry and total disregard for what the show world would make of them. He wanted a true working terrier full of character and courage, and was not bothered if its ears pointed upwards, outwards or even downwards – a black hair in the wrong place was not worth mentioning. If you have the opportunity to study some examples of the Colonel's dogs from the early nineteen-hundreds, you will not fail to be impressed by

The modern West Highland White is a happy and capable dog.

their workmanlike appearance. Some seem to be slightly longer in the back and have a longer tail than modern Westies.

It was taken for granted that the West Highland White would excel at catching rats and rabbits, and by hunting the latter it was expected that its nose and condition would be improved. A tale is told of a lost West Highland White belonging to the Poltalloch estate, as an indicator of its intelligence. It was trapped in a rocky cavern, and daily the keepers would go to the cairn hoping to catch sight of the dog. At last one day they spotted a pair of bright eyes at the bottom of the hole, and the eyes remained visible when the dog's name was called. A quick-thinking keeper wrapped a rabbit skin round a small rock and lowered it by a piece of string to the terrier, which seized upon the idea, grabbed the skin, held on, and was pulled free. Once free, it apparently fainted, but later recovered in the comfort of its own home. It is just this type of quick thinking in a working situation that the rabbiter of today wants to see in his dog.

The West Highland White has much to recommend it to the rabbit hunter: a good grip, protective coat, plucky nature and keen nose fit it for the task. Perhaps one of the most pleasing aspects of the Westie's character is the happiness with which it approaches life, and it would make a cheerful sporting companion. In size, the terrier falls between the lighter Cairn and heavier Scottie, and consequently may be considered a little large by some rabbiters; I must confess that I prefer terriers that are slightly smaller and lighter than Westies. However, I am sure that if given the chance, the West Highland White would gladly work, and these terriers still possess characteristics that make them a credit to their hardworking forbears.

BREEDS OF RABBITING TERRIER FROM THE BORDERS

The Border country where Scotland meets England has been the breeding ground for three distinct terriers, each unique and famed for different reasons: the Dandie Dinmont, the Bedlington and the Border. When these terriers were bred, the Borders was a region of cascading hills clothed with forests; as now, it boasted good farmland and provided an area where otters, foxes and rabbits flourished. In appearance the three terriers are quite different, reflecting the purpose for which they were bred.

The Dandie Dinmont

The Dandie Dinmont, known to some as the 'short-legged tinker's tyke', is probably the oddest-looking dog to be called a terrier. It has what looks like the ears of a hound and the back of a dachshund, and has a tuft of soft silky hair on the top of its head known as a top knot, that it wears like a crown. It also has what is described as 'a voice heavy beyond its weight', which, when taken with these other features, has indicated to many experts that the Dandie Dinmont is the product of crossing a small nondescript terrier with an otterhound. This is preferable to the theory that the Dandie Dinmont is a Dachshund–terrier cross.

The Dandie Dinmont is the only dog to be named after a literary figure, thanks to the popularity of the writer Sir Walter Scott, well known as the author of the story *Ivanhoe*. However, in spite of this rather grand fact, the Dandie Dinmont was bred and kept by generations of gypsies and farmers, and was highly thought of by men who knew the value of working dogs; indeed, so much so that the Earl of Northumberland offered a farm to one of the most picturesque characters to be associated with the breed, Piper Allen by name, in exchange for one of his terriers. The offer was rejected, Piper Allen obviously preferring the itinerant life of a pipe-playing tinker, and the love of his dog, to that of a farmer – for which you have to admire him.

The Dandie Dinmont has much to recommend it as a working companion. It is an ideal height for going to ground, although pictures do show that it was slightly taller during its early years. Its length of back was

The Dandie Dinmont. Ideal height for rabbiting and in possession of a very good grip. (D. Bezzant)

intended to enable it to move with greater ease underground, the theory being that it can curve its body round tight corners that would halt shorter-backed terriers that have less curvature of the spine. According to a writer from the mid-nineteenth century who used the pseudonym 'Stonehenge' and was editor of *The Field* magazine, the Dandie Dinmont made an excellent ratter. Terriers that excel at ratting invariably make good rabbiting dogs, and the Dandie Dinmont has much in its favour when it comes to hunting rabbits.

One advantage that the Dandie has over other terriers is its thirst for, and ability to detect scent, owing to its infusion of hound blood. At a height of between 20 and 28cm (8 and 11in), it can move round the most well

The Dandie Dinmont is a determined and tireless worker but very difficult to acquire.

The Dandie Dinmont has an unusual appearance and was made famous by the writings of Sir Walter Scott. (J.H. Bezzant)

covered warrens; and if there is one thing that the Dandie will never do when it has been assigned a task, it is to give up.

As to his character, it has been said that he may have had acquaintance with knights errant and later crusaders, for when he fights, he fights to right a wrong or to help the distressed. And when he considers his cause is just, it doesn't matter in the least whether his foe be dragon or giant, he will gets his teeth into the hind legs of either. Despite this, he makes a generous friend who expects the same level of generosity from others. The Dandie will not trouble others, and expects others not to trouble it.

The Dandie Dinmont has many strong points, but I do have my concerns as to its suitability for rabbiting. I have no doubt that the breed is full of eagerness to work; however, I feel that it is best suited to subterranean hunting in which it confronts its traditional antagonists, chiefly the fox – which is, after all, what it was intended for. It is lacking just a little in speed and athleticism when compared with some other terriers, and it is not quite fast enough for my liking to be the ideal rabbiting terrier. Rather worryingly, the Dandie Dinmont has a reputation for being aggressive towards ferrets, and would therefore require careful training. It is also known as one of the hardest dogs to breed to comparative perfection, and they can suffer from back problems. In terms of purchasing, you will find there is only a limited number of breeders, and the offspring can be costly.

I'm not sure that the Dandie Dinmont is the ideal dog for the rabbit hunter who has only a limited knowledge of working terriers; but for the experienced handler who wants to prove the working talent of an ancient breed, it could prove to be a good companion.

The Bedlington Terrier

The reader will see instantly that the Bedlington Terrier fails to comply with the model for the ideal rabbiting dog in a number of ways. Mainly it is too tall, and questions have arisen regarding the strength of working instinct that is found in the breed today.

There are, however, valid reasons for its inclusion in our study. People who knew what was required in a dog to succeed at catching rabbits, essentially bred it for this purpose. Secondly, it does not comply with our model because its sphere of work is slightly different to the hunting carried out by other terriers. The model represents those terriers that excel at hunting rabbits at warrens and in areas where they have plenty of cover, while the Bedlington is more likely to catch rabbits as they make a dash over open ground.

The Bedlington owes its name to the mining village of Bedlington, because more of these terriers were kept there than in any other district; however, it might more fittingly have been dubbed the 'Rothbury terrier', for it was from Rothbury forest that it

The Bedlington is a breed build for speed more than for the terrier's traditional work of going to ground.

emerged, thanks to the efforts of tinkers and gypsies. They had set their minds on producing a terrier that would be a prolific pot filler. If they had a dog that failed to perform, not only would they be disappointed, they would have gone hungry and missed out on a much-needed meat ration. Thus they would have put a great deal of careful thought into the features that a rabbit-catching terrier should possess, and there was a large pool of astute terrier men to draw from amongst the bands of itinerant tinkers, whose wanderings included Rothbury forest.

They opted for a terrier built more for speed than earth work, and the height of around 40cm (16in) is not ideally suited to the traditional groundwork of terriers. However, its strength when compared with other

The Bedlington. A tall, fast breed of terrier originally relied upon to catch rabbits but altered significantly due to showing. (D. Bezzant)

terriers that hunt rabbits is its extra pace. I have several times watched my terrier chasing a rabbit in a short sharp dash, and seen it be just lacking, by a yard or less. In such cases the Bedlington would have secured a catch. Originally it had a weather-resistant coat that enabled it to endure working in water, and would stand up to rabbiting in all weathers.

But time has not been particularly kind to the Bedlington, and this was clearly demonstrated by a writer who voiced his concerns about the breed during the 1920s. He was in no doubt that the Bedlington of his day was, as a whole, nowhere near as good as the breed had been forty years prior to his writing. Features such as the length of head suffered from over-exaggeration, and the coat was considered inferior in texture, resulting in the loss of its prized weatherproof quality. Undershot and overshot jaws were far too common an occurrence. By studying early pictures of the Bedlington it can easily be seen how the terrier has changed from a dog that closely resembled a Dandie Dinmont on long legs, to the terrier of today which looks like a lamb.

The modern Bedlington continues to be plagued by problems, the most unfortunate of which is a propensity to retain copper salts in the liver, which has been responsible for chronic sickness and premature death amongst the breed in the past. Reputable breeders have their kennels tested for the recurrence of copper retention amongst their dogs. Some question the working instinct of the pure-bred Bedlington, while others staunchly defend it; consequently it can be a lottery as to the quality of working dog you end up with. A group of admirable characters did attempt to redress this by developing a working Bedlington, and it is the wisest option to contact them if your intention is to use a Bedlington. These men are like the tinkers of old and are trying to make the dog what it used to be, and ought to be now.

It has a very difficult role to fill regarding rabbiting. It cannot compete with the small terrier when it comes to movement around the warren, and it has been surpassed by Whippets and small lurchers, which possess more pace for working open ground. It can still obviously be used to catch rabbits, but it is really a terrier more for the breed enthusiast than the rabbit hunter who is looking for the most useful and best terrier he can get.

The Border Terrier

The last terrier to originate from the Border region is the dog that bears its name. Unlike the Bedlington, the Border Terrier has hardly changed at all from when it first became known, and has continuously been valued for its working prowess throughout each decade of the last century. It may be slightly tall for the perfect rabbiting terrier, but with an impressive height-to-weight ratio and a deep, narrow chest, it is certainly worth a very close look.

The Border was a dog required for a special purpose, and was adapted for work under the most difficult of conditions. The region where this dog was put to work was riddled with peat moss holes; these were very wet and could be dangerously deep. A wet moss hole may be of considerable length, and slits in the rock may be from 6 to 9m (20 to 30ft) deep. One moss hole in the Cheviots had a sinister reputation, based on the loss of at least half a dozen dogs in it. Nothing, however, will deter a Border from pursuing his traditional foes. Although taller than the likes of a Cairn, the Border is still a small-bodied terrier with recorded weights as low as 5.5kg (12lb) for some working examples of the breed. This provided them with the narrowness of body to follow foxes into the rocks. It has been said that only those who have seen can understand how small a slit in the rocks a fox can get through, and it would be impossible for the likes of a fox terrier to follow him.

This ability to negotiate the most impenetrable of natural obstacles is an ideal feature of the breed when it is used to pursue rabbits as well as foxes. Added to which, it has more than enough stamina for a day's work as a

The Border Terrier has remained unchanged since its earliest days as a breed.

result of it being expected to run with the hounds – and when a hunt did manage to run a fox to ground, it would be the Border terrier that would have to follow the fox into its lair, while the hounds enjoyed a period of respite from their exertions. Furthermore, not only will the coat keep out the horrible wet of the moss hole, it will enable the terrier to rabbit quite happily in all weathers. Borders are tough little creatures and have been known to keep at their work despite horrendous injuries, such as a broken jaw.

I found an old article from 1963 concerning Border Terriers, in which the writer pointed

The Border Terrier. A terrier with a steady character and bags full of stamina. Still a very capable worker today. (D. Bezzant)

out that from the end of World War II to the time she was writing, over 150 Borders had acquired working certification, and even now there is a plentiful supply of working Border Terriers. Having endured time without any apparent changes, the Border remains a most capable sporting companion and is immensely useful to the contemporary rabbiter.

BREEDS OF RABBITING TERRIER FROM ENGLAND

There are five breeds of terrier from England that are of particular interest to the rabbiter: the Yorkshire, the Patterdale, the Norfolk and Norwich, and the Manchester terriers; and there is quite a difference in their appearance and history.

The Yorkshire Terrier

It may sound like an amusing prospect to combine a Yorkshire Terrier with a rabbiting man, and the reader may well be asking himself what use such a terrier can possibly be. This is understandable when you consider the ridicule made of the breed within the show ring; but beneath the long coat and behind the absurd bows placed on their heads, there beats the heart of a terrier that has, during its development, been associated with catching rabbits.

Dan Russell, an ex-Fleet Street journalist and lifelong terrier man, mentions in his excellent little book entitled *Working Terriers*, Yorkshire Terriers with regard to hunting rats and rabbits. You may be surprised to hear that what he says is not at all derogatory. He states that when writing in 1948, Yorkshire Terriers were grand little chaps for ratting: they could shift a fox from a drain, and were considered useful for rabbiting. They were so small they could get down an average bury so there was no need to take a ferret. A friend of Russell's had used Yorkshires for many years and would have no other type of dog. His terriers weighed between 2.3 and 3.6kg (5 and 8lb), and were thought of as thorough little sportsmen that were as game as they make them. Dan Russell, who had a wonderful knowledge of working terriers, believed that for the man who likes a little ratting or rabbiting in his spare time, a Yorkshire is a grand companion. They were the only terriers small enough to be used instead of a ferret, and yet strong enough to tackle a big rat or to draw a rabbit if he is not too far in.

Originally the Yorkshire Terrier was a hunting terrier weighing between 4.5 to 5.5kg (10 to 12lb), and the working men of the West Riding of Yorkshire are usually credited with the production of the breed; the Yorkie became known as a constant companion to the industrial classes in and around the region. The working man evidently thought that he needed a much smaller type of terrier than the Airedale, and is presumed to have resorted to crossing this breed with some of the short-legged breeds such as the Skye Terrier. There is also a suggestion that broken-haired terriers already established in the county during the late eighteenth century, were interbred with the Clydesdale terriers brought by Scottish wool workers when they moved to the industrial towns of Bradford, Halifax and Leeds.

The early Yorkies had a specific function to perform, which was to keep the homes clear of rats. This was a vital task requiring speed, determination and manoeuvrability, and the little terrier was up to the job. As time progressed, selective breeding reduced the size of the breed, and they now weigh up to 3kg (7lb); but this by no means renders them useless. Once again, returning to Dan Russell, he is of the opinion that the smaller a Yorkshire Terrier is for ratting, the better, and this is equally the case with rabbiting.

The Yorkshire Terrier is suited to hunting the rabbit around the warren and in those places where the rabbit hides and no other dog can reach. From my own recollection, I can think of a number of situations where the size of these terriers would have been of use, such as amongst woodpiles, in farm buildings and in rocky outcrops. It has sufficient

The Yorkshire terrier, despite the image given to it by show contenders, can be a very useful working terrier and its size makes it particularly suited to some aspects of rabbiting.

strength and determination to deal with a rabbit, and is described as a dog that is 'all terrier' and relishes the chance to show its working skills. It is also an extrovert and affectionate dog that is reputed to be as agile as a grasshopper; it loves to be active, and can jump astonishing heights for its size.

The Yorkie may lack the speed and strength of other terriers, and is not a dog to work the open spaces, but for speed and movement in tight spots they cannot be bettered, and there are plenty of such places where rabbits seek to hide. I regularly hunt rabbits that hide in a woodpile that adjoins a wooden shed, and under the floorboards of a derelict farmhouse where they seek refuge. In these locations space is at a premium and a Yorkshire Terrier could be invaluable. If you hunt in similar situations, I would recommend that you seriously consider a Yorkie; but I would tend to use something else if you hunt on exposed hillsides. Those who wanted a Yorkshire Terrier would do well to look amongst the animal shelters, because I have seen numerous Yorkies that are in need of good homes, and which look well suited to work.

The Patterdale

The first terriers that my brother saw hunting rabbits were a Jack Russell, a West Highland White and a Patterdale. During the hunt the Patterdale got itself stuck in a hollow log, and was grumbling away to itself in much the same way that a person will reproach himself if he does something stupid. Playing the good Samaritan, my brother offered it a helping hand in order to extricate the terrier from its predicament, and nearly got bitten for his trouble. From that day to this, he has not had a good word to say about the breed; but even he cannot deny the fact that Patterdales are true working terriers, and have escaped the refinement and exaggeration of features that has ruined some other breeds of terrier.

Sometimes considered a rival to the Jack Russell in terms of working ability, it should be realized that the Patterdale's objectives when hunting are different, and so too is its working behaviour. This is probably best understood if we consider the environment in and the purpose for which the Patterdale Terrier was first bred.

The Lakeland Terrier. The Lakeland and Patterdale share the same foundation stock. (J.H. Bezzant)

The Patterdale is a native of the Lake District, and its name is taken from a sweeping valley in the heart of the country hunted by the Ullswater Foxhounds. Its working environment lies within the Fells, which rise 600m (2,000ft) above sea level and feel like the Arctic in midwinter. Such an environment will obviously not allow for the mounted hunts found in the south, where the terrier was required to bolt the fox, seen as a sporting antagonist, in order to allow the ridden spectacle to continue. In the fierce north, the huntsman had to walk up steep hillsides and down rocky valleys with his hounds and a pair of Patterdales by his side. The intention was that the hounds would run a fox, viewed as a pest to local shepherds, to ground and the terriers would then bolt or kill it. Facing an opponent of 7.6 to 10kg (17 to 22lb) that is fierce and tenacious of life, and hidden away where the closeness of the rocks offers little room to move, even less to grip on to, is not an easy task and calls for certain attributes to be present in the terrier.

In physical structure it must have a deep narrow front and long legs so as to be able to cover many miles of undulating Fell country in all weathers. The hindquarters and loins must be strong to provide the power to jump and climb on to and off rocky ledges without the advantage of a run-up. And in order to endure many hours exposed to the severe Fell weather, the Patterdale possesses a coat that is hard of texture, dense, wiry and weather-resistant. The Patterdale is a tough, hard-biting terrier with a strong jaw, and it will confront a fox without a second thought. The breed has no lack of courage or inclination.

The Patterdale dates back possibly three hundred years, and has changed little during this time. Its origins are similar to the Lakeland terrier of today, except for the inclusion of Bedlington blood in the Patterdale strain.

The Patterdale is a terrier that has been bred intentionally for work, and not for the sake of appearance, and this makes it a contender as a rabbiting terrier. It is an ideal height and weight, is full of working instinct and as hard as nails; however, through no fault of its own, but because of the nature of the work it performs in its native home, it can be a bit fierce for rabbiting. Its inclination is to kill the prey, and it does not easily become a friend to ferrets; in fact I have known some examples of the breed that have killed their owner's ferrets at the first opportunity. They are best worked alone, or with terriers of their own kind – and even then I have seen them make fools of experienced sportsmen.

Patterdales are extremely tough dogs that do a hard job in an inhospitable environment. Consequently, they are best left to experienced terriermen who are confident that they can train the dog properly and keep its fierce side under control. Physically the breed is ideal for rabbiting, though mentally it is more suited to its traditional foe, the fox.

The Norfolk and Norwich Terriers

Despite having different names, these two dogs show a marked resemblance to one another, which is indicative of the common root they share. They are not dogs of great antiquity, but have been known as sporting companions since the early nineteenth century. Both types were known as Norwich Terriers, and were recognized by the Kennel Club in 1932 as a breed – but contentions soon arose over, of all things, ear carriage. Some wanted their terriers to exhibit pricked ears, which had to occur naturally once cropping was outlawed, while others were quite content with terriers whose ears folded over. Such was the dispute that these terriers could no longer dwell happily side by side with the same name as they had done for at least 130 years, and in 1964 they were subdivided into the Norwich and Norfolk breeds, with pricked and folded ears respectively.

Working enthusiasts were, and continue to be, bemused that such intensity of passion should be devoted to something so insignificant. They knew that both these terriers were accomplished hunting dogs, whether the ears pointed upwards or downwards. However, terrier men did hold opposing views regarding their suitability for going to ground after foxes. What they did all agree on was that both the Norwich and Norfolk were the best of terriers for rabbiting, indeed arguably beyond equal at the sport. It is of interest to us to ask why this was, and what it is about these terriers that suits them so well to the work.

The answers can be found in their background and development. The Norwich, as it was known, found its employment amongst the farmers of East Anglia, a region that has for centuries past favoured the growth of

The Norfolk and Norwich terriers had an enviable reputation as ideal rabbiting dogs, and continue to be a good size for the work.

arable crops, unlike the grassy hills of northern England and Scotland that are suitable only as pasture to support livestock. Quite obviously these farmers confronted different foes, the shepherds of Scotland faced with the havoc caused to their flocks by foxes, and the farmers of East Anglia more worried about rabbits and the way they munched incessantly through their crops – and they were unwilling to sit by and watch their livelihood disappear into a rabbit's stomach.

They did not opt to rely on time-consuming traps and snares, but turned to a little red-haired terrier that was known within the region. The Norwich was popular with the privileged undergraduates of Cambridge University, who used them for the odious purpose of entering rat pits in which the dogs would have to face a company of rats. The dog was required to destroy the rats as quickly as possible, and would receive painful bites from the terrified creatures. The terrier had to be as hard as nails to survive because it could not negotiate a treaty with the rats, or expect a compassionate rescue from its owner, no matter how many yelps of appeal were made. There was money at stake due to the placing of wagers, and it shows that education and decency are not inseparable bedfellows.

The Norwich was bred by horse dealers, horse breakers and dog breeders from within the region, and it rapidly developed a reputation for its ability to catch both rats and rabbits. An answer to the farmer's prayers, the fiery terriers were put to work along the miles of hedgerow that would have existed during that time, and are still a feature of many of the counties that fall within East Anglia. The farmers would have expected the Norwich Terriers to have got on with their task under their own initiative and in small wandering packs to bring the number of rabbits under control. As such, they would have been as valuable as hired workers, and would have continued to be vital figures on the farm right up until the 1950s and the outbreak of myxomatosis.

Consequently, the history of both Norwich and Norfolk terriers is interlinked with rabbit hunting from its earliest records. I lived in East Anglia for a time, and while I admit that it has its attractions, it has to be said that the terrain is not particularly demanding nor the weather harsh. Therefore, terriers that are familiar with hunting in this environment

The Norfolk. Ideal size and coat type for rabbiting but difficult to find. (D. Bezzant)

may well be thought to lack the stamina and hardiness to work in the north of Britain where the conditions and landscape are at their most demanding.

Fortunately, some examples of the breed have been employed in every part of the country, and it is from these experiences that we find an answer. As to stamina, the Norwich has proved itself by keeping pace with horses throughout the course of an entire day, and has met the challenge of covering hill and dale unfamiliar to its place of origin. There are accounts of Norwich Terriers applying their industry in the north-east of Scotland where the climate is certainly one of the most off-putting in the United Kingdom. By so doing, they showed that their little red coats were good enough to afford them protection from all weathers.

We can see already that the Norwich has many of the characteristics highlighted in the model of the ideal rabbiting terrier. It is a little dog of intense activity weighing around 4.5kg (10lb) when in working condition, and was supposed to have been bred to a size that would enable it to work a 15cm (6in) drain. They became specialists at the hedgerow, hunting rabbits, which requires intelligence and a willingness to work together in order to succeed. With a natural affinity for hunting rabbits and the physical attributes for the task, both Norfolk and Norwich terriers have in the past been counted amongst the very best rabbiting terriers to have been bred.

However, from a working perspective, things do not look that good for either the Norfolk or Norwich today, as I realized when I spoke to a leading breeder of these terriers. I discovered that neither strain is a prolific breeder, and often ends up with a litter of no more than three puppies. They are therefore reasonably scarce, and as a result cost a small fortune. I was quoted a price of £800 for a puppy – and it should be remembered that this premium is not due to the possession of an excellent working pedigree; in fact neither the Norwich nor the Norfolk Terrier has been offered the opportunity to work consistently in recent times. However, I do believe that they are still imbued with the keenness and ability, to catch rabbits.

For the sportsman considering the purchase of a Norwich Terrier, he should make sure that it does not come from a line where the dogs have become too heavily built. During the period between the wars, the Norwich nearly became extinct and bull terrier blood, along with some other terriers, was introduced to try and revitalize the breed – this still shows occasionally today in those Norwich Terriers that have too broad a chest and are shaped like a barrel. The Norwich is also a breed prone to epilepsy.

The Norfolk Terrier has remained free from the influence of other breeds, and is very much as it first was with regard to its physical dimensions. However, beneath the healthy façade the Norfolk too has its problem, in the ugly form of liver dysfunction, which will cause sickness and possibly death.

Therefore, both of these terriers face difficulties today and, with such a hefty price tag, are unlikely to be a choice of the rabbiter – which is a shame, because they have all the raw talent to become once again one of the best rabbiting terriers to be seen in the British Isles. I think if somebody with very deep pockets were to give the Norfolk and Norwich terriers a chance, they would quite quickly build a reputation to rival that of their ancestors.

The Manchester Terrier

At 38 to 43cm (15 to 17in) high, the Manchester Terrier is slightly tall for a rabbiting terrier, but its predecessors' reputation as one of the foremost killers of rats makes it worthy of consideration because the skills required for rabbiting are so similar. The Manchester Terrier of today is generally thought to be the product of adding whippet blood to the Old English black-and-tan terriers that have been known in this country for approximately five hundred years, and the Manchester Terrier was itself known as the 'black-and-tan' terrier until the 1920s.

The original black-and-tan terriers would have been an assorted bunch with similar outlines, varying weights and cropped ears, until 1895 when this latter practice was made illegal. The earliest keepers of these dogs would have had no consideration for a breed standard, and gave little thought to whether a dog looked right or not. They wanted only one question answering: 'Can the dog work?' Apparently it could, and it took part in every form of terrier work; but as time passed, the ancestor of the Manchester became much more of an urban dweller than a rural one, and was confronted with the most prolific pest of all towns and cities, the brown rat.

The rat had plenty to feed on and a prolific choice of squalid conditions in which to make its home. Quick-moving, fierce, well hidden and with an army of supporters, the rats were formidable adversaries, and it took a special kind of terrier to hunt them week in and week out. They had to be fast, agile dogs with lightening-quick reactions and an indifference to the pain of a rat bite, or sometimes repeated bites; they would have been able to turn rapidly on the spot, leap in the air, and catch hold of a rat in the blink of an eye. There would have been huge numbers of rats, but even so, it would not have been an easy task to pinpoint their location accurately amongst all the stinking rubbish, and terriers that hunted rats regularly developed a fine sense of smell. The black-and-tan terrier was therefore regarded as something of a rat-catching specialist.

It was inevitable that these dogs would encounter the darker side of life in the horrendous rat pits where they were expected to kill as many rats as they could in as short a time as possible, and all within enclosures that were usually circular in shape and surrounded by a crowd of eager spectators. Thankfully this 'sport' was outlawed a couple of years before the outbreak of World War I, but a number of Manchester-type terriers had gained notoriety as so-called champions of the pits.

During the 1800s, a man by the name of John Hulme is thought to have combined the black-and-tan and the Whippet to produce the lithe, agile and powerful dog we know today as the Manchester Terrier. Being descended from these two breeds that were synonymous with rabbiting, it is hardly surprising that the Manchester was also considered a capable rabbiting terrier. Even today its advocates claim that it is a good old-fashioned sort of dog, unspoilt by the show ring and still an adept hunter of rats and rabbits.

Its critics say that it has got too large and has lost its working instinct, but these comments are based on its suitability for the traditional terrier work of going to ground, which the Manchester Terrier may not be willing to do. However, these views should not be applied to the terrier's willingness to hunt rabbits, because very few dogs are not excited by the sight and smell of rabbits. Admittedly, however, there are several accounts suggesting that this terrier will not confront any hard-biting foe, and thus lacks 'gameness'.

Nevertheless, I believe there is still enough about the Manchester Terrier to make it an able hunter of rabbits today. Its sleek, smart coat is not the most weather-resistant, but it does have the advantage of not picking up mud and dirt, which was of vital importance when the terrier had to search the street rubbish in pursuit of a rat. It may be a little large; however, as compensation it is a fast-moving terrier that will excel at catching those rabbits that break cover as they make a dash for a better hiding place.

The Manchester Terrier is not a popular breed, and is therefore quite difficult to acquire. It is best suited to someone who is confident that he can bring this terrier's latent working talent to the fore.

A RABBITING TERRIER FROM IRELAND

Having carefully listed the attributes required in a rabbiting terrier and drawn up a specific model, you might well wonder why I am writing about a terrier that so obviously contradicts many of my earlier comments. However, although I would never claim that the Irish Terrier will make the ideal or best all-purpose rabbiting dog, it still does have a lot to offer.

Many of the terriers of Ireland were bred with a number of purposes in mind, instead of just one, and the terrier that bears the name of its country is no exception. It would have been expected to:

- catch rats on land and in the water;
- hunt the hedgerows and catch rabbits in the open; and
- guard the home against intruders.

Time and again during my reading I came across the terms 'excellent rabbiter' and 'ideal sporting companion' spoken of in conjunction with this terrier. This prompted me to take a closer look at the dog, which is holder of the alarming title of 'daredevil' amongst the terrier group.

Like most other terriers, there is nothing exact concerning the origin of the Irish Terrier, although the romantic notion has been put forwards that it has existed as long as that country has been an island. It is generally believed that the black- or grizzle-and-tan terrier of Britain was the foundation stock for the Irish as well as the Welsh and Scottish terriers. The recurrence of black-and-tan puppies in a litter of Irish Terriers is offered in support of this.

Early examples could be found weighing a miniscule 4kg (9lb) or less, and possessed either black-and-tan, grizzle-and-tan, wheaten, or any shade of red coat. During the 1870s the minimum weight requirements for Irish Terrier show entrants gradually increased from 4kg to 5.5kg (9lb to 12lb) and then to 7.25kg (16lb). Following the initiative of Belfast breeders to form the Irish Terrier Club in 1879, disputes about the most desirable weight became a common occurrence. Doubtless there were concerns that increases in weight would compromise the working abilities of the terrier. Finally, the weight was fixed at 12kg (27lb) for dogs and 11.25kg (25lb) for bitches, and red of a deep auburn shade became the dominant coat colour. By the 1890s a much more standardized breed had emerged, approximating the popular Irish Terrier of today.

An Irish Terrier, although tall, is a terrier with speed and has maintained a deep, narrow chest.

Irish Terriers were bred and kept for work, and many believe that as a sporting companion there is no equal to the Irish Terrier. They are extremely hardy, soon learn to love a gun, and can easily be taught to retrieve from either land or water. They possess an excellent nose and will readily slot into the role of a spaniel. However, they have not lost an iota of their true terrier instinct, and are clever at marking rats or rabbits in their holes and will face fox or, historically, badger. The Irish Terrier is quick to mature and unquestionably game: a three-month-old puppy that killed a full-grown rat is testimony to that!

Mrs Edna Howard Jones, a championship show judge and successful breeder and exhibitor, also claimed that the Irish Terrier excels at catching rabbits. She recalls watching two of her bitches work a hedge, and described it as a picture of intelligence and co-operation. One of the terriers waited silently, pointing on one side, while the other, utterly regardless of thorn and bramble, would work the hedge and drive the rabbits with great speed. She also points out that those Irish Terriers that were brought up during the years when myxomatosis cleared the landscape of rabbits, never lost their working instinct, and when rabbits returned to the hedges and fields near her home, were able to hunt them as skilfully as if they had done it all their lives.

The Irish Terrier. The Irish is not an ideal rabbiting terrier but is a very capable all-rounder. (J.H. Bezzant)

At just 15cm (6in) below the Airedale, the Irish Terrier is too tall to be the ideal rabbiting terrier, but it does possess a good outline, with depth as opposed to breadth of body. They will, however, obviously have limitations as to where they can go after rabbits, a state of affairs that does not apply to many of the other terriers used for rabbiting, which are much shorter.

In its favour, the Irish Terrier is active and has a good turn of pace. It has an instinctive feel for catching rabbits, and is relatively easy to train, as military trainers during World War II discovered. It is credited with being more anxious to please its owner than any other terrier.

The Irish Terrier is more of an all-purpose dog than a champion of one activity, and is ideal for the person who wants a dog that he can take rabbiting one day and shooting the next, and which is also capable of guarding the family.

A RABBITING TERRIER FROM WALES

I used to live in south-west Wales, and visited Sealyham mansion, in Pembrokeshire, where the terrier that shares this name was bred. It was easy for me to imagine what a grand sporting estate this property would have been in the mid-nineteenth century when the founder of the breed, an eccentric aristocrat named John Owen Tucker Edwardes, is most likely to have started his work of moulding the Sealyham. Within twenty-four years of its inception, this terrier had become one of the most popular breeds of the day, all thanks to its working ability. So what had Tucker Edwardes managed to achieve that made his terriers more popular than other breeds, and what prompted him to develop his own breed?

In his day, the county of Pembrokeshire was considered to be a remote and isolated place, so much so that well-to-do families were said to have only limited choice: either they consumed large amounts of alcohol, or they took part in field sports for their entertainment. John Owen Tucker Edwardes chose the latter, and for good reason – the River Sealy, on which otter were to be found, passed within a stone's throw of Sealyham, and the valley where his house nestled supported a multitude of badgers, foxes and rabbits. In his forties by the mid-nineteenth century, he would already have acquired a wealth of sporting experience and would have known the estate, which had been in the family for several hundred years, like the back of his hand. Just like sportsmen of today, his years of hunting would have provided him with firm opinions regarding the standard for a working terrier which, at that time, would have to be sufficiently versatile to deal with the native otters, badgers, foxes, rabbits and rats. For some reason he was not totally satisfied with the type of terrier he found to be common within the county, and was not inclined to import a terrier breed from the neighbouring countries. This was probably because he did not like the way many of the registered breeds were developing; furthermore, if living on the fringes of Britain was anything like it is today, it would have been frustratingly awkward to import things, and presumably this would include terriers.

John Owen Tucker Edwardes was confident that he could produce a terrier to suit his needs from the raw material found within the county, although no one knows exactly what dogs were used for the breeding of the Sealyham. He was not of a disposition to make records of his endeavours, and obviously believed that the limbs and body of the Sealyham were more eloquent than pages full of written remarks.

His requirements were that the terrier be small, game, active, agile, weighing between

The Sealyham is a Welsh sporting aristocrat's idea of a perfect working terrier.

6 and 7kg (14 to 16lb) and 20 to 30cm (8 to 12in) in height. His intention was to have a terrier that would hunt with his otterhounds – and by all accounts, he succeeded. The Sealyham was recognized as a sporting terrier in the truest sense of the word, and quickly became a favourite with the other notable families to be found within Pembrokeshire. One of the most tireless advocates of the breed was Captain Jack Howells of Trewelwell, near St Davids. He was Master of the Pembroke Hunt, and used a pack of Sealyhams to hunt otters on the River Solva. He eulogized the intelligence and mobility of the breed, and pointed out their talent for driving rabbits from the cover or gorse (that he called 'furze').

There were critics of the breed who complained that the terrier was rarely seen outside its native shire, but there were just as many visiting sportsmen to the county who were impressed with what they saw of the Sealyham at work. Furthermore the exploits of Sir Jocelyn Lucas MP supply ample evidence that John Owen Tucker Edwardes had managed to produce the capable working terrier he had dreamed of. Jocelyn Lucas was one of the foremost terrier men of the last century, and used his Sealyhams to successfully hunt all manner of legitimate prey at home and abroad; indeed, he was so convinced of their working prowess that he placed an advertisement in *The Times* personal column in August 1921, challenging the owners of any other breed or pedigree to an open contest with his Sealyhams, whether it was underground to badger, in water to otter, or on land to stoat, rats and rabbits. Not a single person accepted his challenge, which confirmed to him the towering supremacy of the Sealyham as an all-round terrier.

Clearly, the Sealyham that Lucas used during the first half of the twentieth century would have been the ideal rabbiting terrier – but what about the breed of today? I am afraid that Tucker Edwardes would have been disappointed, and from a sportsman's point of view the Sealyham is regarded nowadays as heavily set and a little clumsy. It is commonly accepted that in the period between the world wars, the terrier was made into a far more glamorous dog and lost a great deal of its value as a working terrier. Even Jocelyn Lucas recognized the changes

The Sealyham. At one time, one of the most popular working terriers but became too heavy for many sportsman's liking. (D. Bezzant)

occurring in the Sealyham, and this prompted him to experiment with his own breed of terrier. Working enthusiasts believed that the breed was acquiring too much weight, and opted for smaller individuals that were sometimes available, because they thought these were better adapted to sport, being more agile and quicker in movement. It should be remembered that they wanted the Sealyham to go to ground, and because the changes arising from selective breeding effectively precluded it from doing so, its working credentials were therefore thrown into doubt.

However, for the rabbiting enthusiast, the ability of the terrier to go to ground is not of prime importance because ferrets could be used to bolt a rabbit. Consequently, we have to ask whether the modern Sealyham Terrier has the makings of a good rabbiting dog. A leading breeder suggested that the answer is yes, and has observed a number of incidents involving Sealyhams in support of her view. Certain individuals amongst her prize-winning Sealyhams would make mock of her fencing and escape to the surrounding countryside, where they would spend many hours hunting and catching rabbits; indeed, two of them used to abscond for days at a time and live on their wits. This shows that, while their inclination for other forms of terrier work may be lacking, the Sealyham is nevertheless keen to go rabbiting. Prior to myxomatosis it would have been quite easy for a dog of virtually any description to catch rabbits because they were so prolific; nowadays, however, you can reside in a rural area and hardly ever clap eyes on a rabbit. Therefore, any dog that is able to catch rabbits in these circumstances without assistance obviously has a talent for the work, and the two aforementioned Sealyhams showed that they were sufficiently intelligent to work together and survive without any loss of condition.

When I was at primary school, my family owned a Sealyham Terrier that hated collies: a collie had once aggressively attacked her in front of our house, and for ever after she blamed the entire breed for this transgression. On that occasion the Sealyham would not back down, and her white coat was quickly covered in blood. My mother feared she was badly injured and rushed her to the kitchen, but as she washed away the blood, she realized that none of it belonged to the terrier. I do not want to give the impression that they are aggressive dogs, but merely to point out that they will stand their ground and can be very determined. Tucker Edwardes always had a group of Sealyhams in his company and they never caused trouble whether they were in the town or the country, or sitting at his feet while he served as a Justice of the Peace.

My major concern about the Sealyham is the strength – or weakness – of its back, which is prone to injury; our Sealyham had to be put to sleep as the result of such an injury, and I have thought that this is a weakness of the modern breed. The Sealyham has not been in the regular employment of hunters for at least sixty years, and it is unlikely that you will be able to find a puppy that has working parents. At the time of writing, the cost of a puppy from a reputable breeder averages £600. These terriers have changed a lot since their early days, and cannot compete with their predecessors in terms of sporting prowess. Nevertheless, the modern Sealyham still has a lot to offer, and I believe that regular exposure to rabbiting would soon rekindle its working instinct and capabilities.

A RABBITING TERRIER FROM AUSTRALIA

It is not surprising that, in a land that has seen the development of regional terriers during the previous few hundred years, we should give little thought to terriers emerging from other countries. However, one non-British terrier that has won a small group of admirers, due to both its working abilities and its tenacious character, is the Australian Terrier. It has been described as a game little dog, possessing much of the spirit of its early ancestors.

Although Australia may have been the land that shaped this little terrier, the rootstock from which it sprang definitely hailed from our shores. If you study the Australian Terrier carefully, you can see clearly that there is a strong Scottish influence in its make-up, most obviously from the Cairn and Skye terriers. The Australian Terrier emerged when Scottish economic migrants took their hardy little terriers with them when they left their homeland in favour of a new life in Australia. The land of Australia was unlike their native Scotland, with the most notable differences being the terrain and weather, and so the way they were farming would have to change. As you would expect, they had to contend with numerous creatures, some of which they would have been familiar with and some they had never seen before. The terriers they had brought with them had done a superb job back in Scotland, but now they needed to change to meet the demands of this new environment. This was done by cross-breeding their terriers with one another, and anything else they could find to throw into the pot: Jack Russells, Patterdales, Yorkies and who knows what else would have been included. The aim of all this cross-breeding was to produce a lighter, quicker terrier whilst retaining the hardiness and working abilities of the Scottish breeds.

As we look at what they wanted this new terrier to do, we begin to understand why they wanted something small, light and fast. There were rats and rabbits to deal with as usual, but this new continent also presented an unfamiliar creature that drove terror into the hearts of the new arrivals – the snake. Remember that we are looking at a time prior to advanced medicine and before the existence of anti-venom, and not many would survive a poisonous snake bite. Men from Scotland and England would not have known one snake from another; to them, all snakes would have been considered dangerous, and if one came too close to them it had to die.

Contrary to the clips in many Westerns, something as narrow and as fast as a snake is not easy to shoot – most people have trouble trying to hit even something as big as a pheasant. And as anybody knows who has ever tried to hit a rat with a shovel, it usually takes a number of blows before you strike your target, and I am sure that you would not want to be engaged in such a hit-and-miss activity if the rat had a bite containing deadly venom like that of a snake. Thus, the new arrivals in Australia had to find a successful way of killing snakes, which was efficient and did not put them at risk. For this they turned to their terriers, but the ones that they had were too big, which is why they reduced the height and weight of the dogs they had brought from their homeland. Believe it or not, the new scruffy terrier that they developed, which could be found on every stockyard and holding in the outback, was an extraordinary snake killer, and probably saved a substantial number of lives. The Australian Terrier was so good at killing snakes that many of them were sent to India to do the same job.

One of the other jobs that the Australian Terrier took on was the protection of the gold mines. The history of this is not clear, but the terriers may actually have been kept in the mine and accompanied the miner when he went to dig for gold. As you probably already know, terriers are very good at giving voice when a stranger approaches their domain, and the Australian Terrier is one of the best at this. He will make enough noise to waken the dead, and is so sensitive regarding scent, vision and hearing, that nothing will get past him. So why didn't the miners select a much bigger dog to guard their mines? A bulky great guard dog may have been able to protect them from intruders, but it would not have been able to protect his owner from a snake bite. A single terrier may not be that formidable, but the miners would have kept a number of them, which is a different matter altogether. Half a dozen terriers attacking as one from six different directions, with six sets of teeth, makes for a very severe deterrent.

The Australian Terrier was also used in what can only be described as the war against the rabbit. The rabbit was introduced to Australia by the early settlers and spread as rapidly and dramatically as wildfire, and caused nearly as much devastation. To illustrate just how quickly the rabbit spread across the country, we can refer to the account of Mr Thomas Austin who owned an estate in Victoria. In 1859 he imported twenty-four wild British rabbits, which he released on his estate for the sport they would provide. Austin was soon able to hunt them, and within only six years, 22,000 rabbits had been shot on his estate – and yet this hardly dented the population.

The terrier from Australia is active, intelligent and ideally built for rabbiting.

Furthermore, Austin's rabbits had spread from his estate to Queensland, which was nearly 800km (500 miles) away, taking on the nature of a biblical plague as they swarmed across the land. Every means was engaged to tackle this rapacious enemy that was consuming the countryside: guns and traps were used, and of course the Australian terrier, who is a marvellous rabbit catcher thanks to his speed, size and tenacity. The rabbit problem was so serious that the authorities decided to build a fence across the country to try and keep the rabbits out of the areas they had not colonized. One of these fences, simply named Fence Number One, stretched for over 1,600km (1,000 miles), and took a thousand men to build; all the materials had to be carried on pack animals, which included camels, and it took five years to complete. This massive construction cost a small fortune – and proved to be useless at halting the spread of the rabbit. Without doubt, the scruffy little Australian Terrier had more of an impact on the rabbit than all man's costly efforts with his thousands of miles of fencing.

As a working dog, the Australian Terrier has much to recommend it. It can take an active part in any form of legitimate terrier work, although it is best suited for rabbiting and ratting. It will go to ground readily, and many will bolt a fox. The early Australian broken-haired terriers were renowned for their sporting abilities, and it would appear that their offspring of today are equally capable of distinguishing themselves. In fact, the owners of these terriers claim that their dogs possess all the attributes of the so-called Jack Russells.

Added to this, the Australian Terrier is a hardy dog, capable of withstanding any weather or hardship, and is as determined and plucky as they come. Loyal and affectionate, they make a pleasant, no-nonsense companion for the sportsman who does not want a fussy dog that barks a lot. In spite of this, the Australian Terrier remains quite a rare breed within the United Kingdom, particularly in working terms, and they are difficult to get hold of. They are not well known, largely because there are not that many of them, and probably because those that are seen are mistaken for other terriers. It is also worth remembering that, in order to secure a following in this country, they have to compete with generations of proven and fondly thought of British terriers. However, should the rabbiting man decide to track down one of these terriers, he will acquire a dog that is of an ideal height, weight and temperament for rabbiting. The Australian Terrier also has the stamina and coat type to undertake strenuous activity in all weathers.

THE DACHSHUND AS A RABBITING DOG

Mostly thought of today as a pet, the Dachshund is a dog of remarkable sporting talent with an indisputable working background. Immediately identifiable – the breed was once said to be 'measured by the yard' due to its long, low body set on short, crooked legs – it has an even, calm temperament and is as keen as mustard when working. Able to perform the task of terrier or hound with equal alacrity, the Dachshund sounds too good to be true, and were it not for accurate records, its endeavours would not be believed but quickly brushed off as just another of the imaginative sportsman's yarns.

Exactly where the little dog came from is hidden in obscurity and is consequently a matter of considerable debate. Some enthusiasts suggest that it was a contemporary of the Pharaohs; this is due to the inscription of dogs resembling Dachshunds on the tombs of Antifaa II near Thebes and on Thotmes III at Beni Hasan. Others disagree, claiming that these portraits could just as easily be a whole host of breeds.

However, it seems the Dachshund has been in existence for four hundred years, based on a variety of drawings, woodcarvings and woodcuts from the sixteenth century. The Dachshund has been resident in Germany, the land with which most of us associate them, for at least the last two hundred years. It is thought that refugees from the French Revolution took their hounds with them to Germany where gamekeepers made use of them. Even during the early part of the nineteenth century, Dachshunds enjoyed a privileged relationship with the German aristocracy, adorning their grand homes and working on their large estates.

The Wire-Haired Dachshund is the favoured type of the sporting enthusiast and is considered more terrier-like than the other Dachshunds.

True to the literal translation of their name, Dachshunds were employed as subterranean badger dogs, as well as for driving game and tracking wounded quarry. Thanks to Prince Albert, the dachshund first appeared in this country during the late 1840s; the dogs he brought over were bred by Prince Edward of Saxe Weimar. It did not take long for the breed to become popular, but it remained difficult to acquire a dog of real quality unless you were the lucky recipient of a gift from the German nobility. But by 1881 this barrier was breached, and this country's first Dachshund club was formed. As the number of breeders increased in this country they became divided over the contentious question – is the Dachshund a hound or a terrier? This difference in opinion was reflected geographically, as those in the north favoured a lighter terrier type and those in the south insisted that the Dachshund was a hound, and promoted a heavy body and a head like a foxhound. In time the north prevailed, and the smaller, neater dogs became the accepted standard. The most terrier-like Dachshund, the Wire-Haired Terrier, appeared after 1914. Its characteristics were largely attributed to the belief that the Dandie Dinmont blood was used to produce the distinctive coat. Some thirteen years later the Long-Haired Dachshund was shown in this country for the first time.

With its strong working instinct to recommend it, proven time and again during the preceding two hundred years, the most obvious progression for the new devotees of the breed was to form working Dachshund clubs. It may be justly asked why, in this land of terriers, the need was seen for yet another earth dog, but we are quickly reminded of the equal combination of terrier and hound qualities that are contained in this little long dog.

The well respected author of canine books, Robert Leighton, writing in 1922, was glowingly enthusiastic over the adaptability of the breed made possible by the possession of the following characteristics: 'a wonderful nose, remarkable steadiness, ability to work out the coldest scent, good voice when on line, incredible pace over ground, iron will, willingness to go to ground, combined with trainability and absence of terrier mischief.'

Working dogs were expected to negotiate unhesitatingly small obstacles, to leap

Mrs Wendy Annette Riley with some of her working Dachshunds during the late 1940s.

brooks, to face water readily and to complete a distance of between 13 and 16km (8 and 10 miles) across country. According to Wendy Annette Riley, founder of a working Dachshund club during the 1940s, club members also wanted a dog they could thoroughly train to stock and poultry, and which was not aggressive towards other pack dogs. When properly brought up and exercised together, cases of fighting were unknown among her dogs. It was also important that they hunted well together, rather than each dog going off on its own line, that they were responsive to the voice and went forwards with courage.

Working Dachshund clubs encourage any form of legitimate sport from rats, rabbits, hares and foxes, to shooting. In spite of their peculiar stature, many Dachshunds excelled as gundogs. They are also excellent swimmers, and Leighton suggested they would have made useful assistants to otterhounds. Wendy Annette Riley in 1948 stated that 'to see a pack of Dachshunds going across country, is like watching ripples and small waves.' Although Dachshunds love horses, Riley felt that foot hunting was best. She found that her dachshunds proved to be businesslike, tenacious, bold, untiring and essentially honest, with such an accurate nose that time was rarely wasted on an uninhabited earth or warren. Consequently, she enjoyed ongoing success in dealing with a variety of quarry with her unusual small pack.

Further examples of Dachshunds being employed with effect were documented by Jane Buckland in her article on the breed in *The Field* in 1959. She mentions Mr Rennie Hoare, who hunted a pack of Dachshunds on his estate. With four couples, he successfully exterminated the rabbits on his estate before the onset of myxomatosis. However, the jewel in the crown of working with Dachshunds was probably the outstanding 'Tencombe Manka', who held a working certificate from the South Berks Hunt and retired at the age of eleven after a full working life. He regularly went to ground in turn with the hunt terriers, and was also a first-class gundog. Even owners of pet Dachshunds unwittingly witnessed the manifestation of their inherent skills as their dogs killed rats, rabbits and other vermin, also taking every opportunity of going to ground.

Today there are six recognized varieties of Dachshund: the Long-Haired, Wire-Haired and Smooth-Haired in standard size and miniature. The question of whether the Dachshund is a hound or a terrier remains unresolved, because there appears to be too much of each for a definitive answer to be agreed upon. Leighton stated that the dog's wonderful powers of scent, long pendulous ears and, for his size, enormous bone structure speak of his descent from hounds that hunt by scent; while his small stature, iron heart and willingness to enter the earth bespeak the terrier heritage.

You will gather from what I have written that the Dachshund is an accomplished sporting dog, and it would appear that the modern breed has not lost any of its working instincts.

Both standard and miniature Dachshunds are capable of being effective rabbit hunters, although the miniature arguably has more to offer. The miniature Dachshund was actually developed by German sportsmen during the late nineteenth century, their intention being to produce a breed that could go to ground after small quarry such as rabbits. Weighing not more than 5kg (11lb), the miniature Dachshund was ideal, and has rabbit hunting in its blood. Although only small, they are active dogs and fully able to endure the rigours of a day's hunt. The Dachshund will be a clever worker at the warren, and excels at picking up and following the trail of a rabbit. The only problem with the Dachshund is the length of back peculiar to the breed, which is susceptible to injury. But despite the ups and downs of a long show career, often in the hands of people totally uninterested in any form of sport, Dachshunds have retained their nose and working abilities and, if given the chance, make excellent rabbiting dogs.

CHAPTER FOUR

Working the Rabbiting Terrier

Before we look at where, when and how to work the rabbiting terrier, it will be as well to answer the question of why anyone in our contemporary society should deem it as either necessary, profitable or enjoyable to hunt and catch rabbits with a terrier.

IS THE RABBIT A PEST?

While I am reluctant to classify any animal with the rather one-dimensional term of 'pest', it does reflect the annoyance and dismay that the rabbit can create for farmer and gardener alike. It has the ability to multiply rapidly, and is a most inefficient and indiscriminate grazing animal when compared with farm livestock such as sheep. It can cause havoc to the most well-protected garden, and will eat the prized plants with delight.

During the period between the world wars, the rabbit was credited with causing forty to fifty million pounds worth of damage a year to British agriculture. In contrast to this, the income generated from catching rabbits was estimated to be approximately two million pounds a year, but this excluded the trade in by-products such as fur and felt carried on by

Terriers like this have been used for centuries to catch rabbits, and are the traditional and natural way to control this pest.

manufacturers. Although rabbit numbers are a tiny proportion of what they were prior to 1953, the year that myxomatosis arrived in Britain, agriculture still resists suffering the losses and crop damage that rabbits exact. Nobody ever questions that there was, and is, a need to control rabbits.

Dogs, ferrets and nets are amongst the oldest methods used to control the number of rabbits, and although man has also sunk so low as to deploy biological and chemical means in an attempt to radically reduce the rabbit population, the obstinate rabbit, in spite of the horrendous consequences of these methods, has weathered the storm and is once again appearing in reasonably widespread numbers throughout the country. People are in no doubt that it is still necessary to manage the increase of rabbits, particularly in agricultural areas; however, there are plenty of views about the best way to achieve this.

A HUMANE WAY TO CONTROL THE RABBIT

I believe that using field sports to control what are termed 'nuisance animals' is far better than relying on modern pest-control techniques. For example, the traditional field-sport world has given us hawks, ferrets, nets and dogs to use, whereas pest control offers traps, poisons and gases.

The sporting person has always had an interest, albeit maybe a selfish one, in preserving the animal he hunts, otherwise his hunting exploits would quite obviously come to an end. Most field sportsmen with whom I have shared an acquaintance do genuinely possess an understanding of, and respect for, the animals they hunt, and this is by no means a new phenomenon. During the 1950s it was the Masters of Otterhounds who raised alarms about the reduction in the number of otters, and for a long time wildfowlers have led the way in combining hunting with conservation of wildlife and the environment.

Alternatively, pest control has given, with regard to management, myxomatosis and gas. These are both aimed at the destruction of the rabbit, and both leave the meat useless and unfit for human consumption. According to the most experienced and well-respected gamekeepers and countrymen, the traditional method of using ferrets with purse nets and dogs is, when correctly applied, the most effective and sporting way to control rabbits. The rabbits that are killed by this process can be safely served up as food for people and dogs.

The rabbiting terrier is an invaluable member of the ferreting team, thanks to the many tasks it undertakes; whenever I go ferreting, I take my Jack Russell along. He always proves his worth, and in some cases he has been more instrumental in catching rabbits that the ferrets. Sometimes he will catch and kill the rabbits himself, and anyone who has seen a terrier kill a rabbit will tell you how quickly and effortlessly it can achieve this. It would be foolish to make the claim that there is not an element of cruelty involved in this form of hunting, because by its very nature, the killing of an animal is essentially cruel. However, when the killing avoids prolonged and unnecessary suffering, it can quite rightly be said to be humane. My Jack Russell, as a typical example of a working terrier, is a proficient killer of small animals, and whether it be a rat or rabbit, he will not hesitate in his actions and has the knack of knowing how to kill very quickly.

When I lived in Wales, I knew and observed a licensed slaughterman; he had been about his profession for a number of decades and took an obvious pride in the skills he had developed. Although he was a bit of a macabre figure and I would certainly *not* like to do his job, I had to acknowledge and admire how little distress, if any, he caused to the livestock he slaughtered. When it comes to the killing of rats and rabbits, the working terrier can justifiably be defended as a humane practitioner, like my friend the slaughterman, because it is in possession of both skill and confidence.

My terrier is always practising his hunting, and here he can be seen amongst some old car tyres and wheels chasing after mice that are hiding in an adjacent woodpile.

NATURAL AND TRADITIONAL RABBITING THAT BENEFITS THE DOG

Having pointed out most importantly that the rabbit is an animal in need of control and that the working terrier is an animal capable of achieving this in a humane way, we can now consider some other factors that may influence the dog.

Working terriers, just like ferrets, have been used in this country for hundreds of years, and their employment in the pursuit of the rabbit is a rural tradition well worth preserving. Such traditions have been instrumental in shaping the British countryside and the people who live in it; however, contemporary working patterns and alternative leisure interests have dulled the enthusiasm for activities such as rabbiting. It is not uncommon to find yourself in the midst of the country and amongst a small minority who maintain an active interest in field sports. You need only talk to people or open a newspaper to observe how things have changed over the last fifty or so years. At that time you could not imagine people living in the country complaining about either the sound of a cockerel or the smell of manure as they are apt to do today.

It is because of this changing character of the countryside that I am a firm advocate for keeping rural traditions, including field sports, alive lest the skills and knowledge that they require be lost. By today's example, you would never believe that rabbit catching with dogs and ferrets, next to fishing, used to be the favourite pastime of children growing up between the wars.

I am also concerned about what will be used to replace field sports if they stop being practised. For example, the rat catcher of a hundred years ago would have relied upon dogs, ferrets, traps and the gun. Nowadays he relies almost entirely upon poison. In similar fashion, the tradition of badger digging, not to be confused with badger baiting, resulted predominantly in the relocation of badgers from sites where they were not wanted to locations where they could live happily without causing offence to anyone. Today, the badger faces the threat of being gassed due to its wanderings amid cattle. In my opinion, the traditional methods of control are more natural in their application and have the added benefit of neither destroying nor desolating the living environment of the hunted animal.

Britain has a long tradition of using dogs effectively, and this is especially so in the

True sportsmen have an unusual regard for the animals they hunt, and quite often will help the injured or orphaned. Even this hunt terrier appears sympathetic to the plight of these cubs.

case of the indigenous terriers. The tradition of using terriers to catch rabbits is one that we should do all in our power to safeguard and uphold. I have not yet come across a rabbiting terrier that does not thoroughly enjoy its work. My Jack Russell is never happier than when he is searching out a hedgerow, marking a warren or rooting through rocks, and the muckier he gets, the more he enjoys it. Using terriers for work on a regular basis is good for them, both mentally and physically, because it allows them to employ their active mind and natural instincts in a way that no other activity can. Also, just as a heavy horse improves in condition and musculature as a result of doing the job for which it was intended, so too the working terrier is in prime physical condition. It seems ridiculous to me to breed a dog with a specific task in mind, and then not use it for that purpose at every opportunity: valuing a dog for its ability to work invariably results in healthier dogs and better breeding.

In conclusion, I would suggest that the effective employment of rabbiting terriers will go a long way towards satisfying the demands of landowners for rabbit control, it will carry out this control in a manner that carefully considers the long-term effects it has on the rabbits, and it allows the terrier to put to work those instincts that previous generations have been at great pains to develop and preserve.

PREREQUISITES FOR RABBIT HUNTING WITH TERRIERS

Having acquired a terrier, the tendency is to want to get on with actually catching rabbits, and to pick up the tricks of the trade as you go along. There is absolutely nothing wrong with wanting to learn by experience – and there is no replacement for it, either – but it is incredibly foolish to think that you can go hunting without attending to certain details first. These are neither particularly difficult nor irksome to accomplish, and yet they have many benefits, which quickly become obvious. These prerequisites are: the fitness and obedience of the terriers; having the permission of the landowners; and knowing the lie of the land and the habits of the indigenous rabbits.

The Fitness of the Terrier

The first thing I would recommend is that the handler makes sure that his dog is fit enough to stand a day's work. I shall look in some detail in a later chapter at the factors that indicate the working terrier's fitness, and what the owner can do to promote the required strength and vigour. It should never be underestimated how much physical energy and mental effort the rabbiting terrier, with all its senses in overdrive, will use up whilst hunting.

Ensuring that the terrier is healthy enough for hunting will depend to a large extent on the provision of an appropriate diet and regular amounts of exercise. In addition to physical exertion, the owner should give some thought to placing his dog in new environments or presenting it with different or unusual tasks in order to offer it continual mental stimulation. The result will be a terrier that possesses stamina, is well muscled, and has an alert and active mind. I have observed that most terriers are by nature fit and industrious – keeping them healthy is a simple endeavour.

My terrier is fortunate to have unlimited access to 2 hectares (5 acres) of field where he can do exactly as he pleases, and he has the company of other dogs in his wanderings. Spending most of his time outside with plenty to interest him keeps him in prime condition and on his toes.

The Obedience of the Rabbiting Terrier

A terrier that is not at the very least well versed in basic obedience will prove to be nothing but a liability if it is taken out rabbiting. An undisciplined terrier is unlikely to listen to its owner in normal situations and will not pay heed to any instruction whatsoever when its blood is up due to the excitement of chasing rabbits. The dog will be a constant nuisance as it distracts the owner from hunting in earnest and at times puts its own safety in jeopardy. I have seen a dog put to work that was not trained in any way, and the result was a lot of irate shouting and gesticulation by the owner, and hectic running to and fro by the terrier.

The willingness to obey simple commands such as 'come', 'heel', 'down' and 'stay' will make a world of difference, and the handler will be able to relax and enjoy his sport. It is neither fair on the terrier nor wise for the handler to take the dog hunting before it has had the opportunity to master elementary obedience.

Likewise, if the intention is to use the combined efforts of dogs and ferrets, they should have been given the chance to develop a sufficient level of familiarity so that they recognize one another as allies and not enemies. On no account should a terrier's first view of a ferret be on the day of a hunt and at the site of a rabbit warren.

The Permission of Landowners

Few of us are lucky enough to own enough rabbit-populated land to support the weekly practice of the sport of rabbiting throughout the winter; we are obliged to rely on the goodwill of those who own farms and smallholdings and other such grounds. Fortunately, many of these owners are quite keen for rabbits to be controlled, but are not interested in pursuing the activity themselves and consequently welcome those who offer to catch the rabbits, especially when it is for free.

When seeking permission, you should always explain how you intend to use the terrier, and ensure that the landowner has no objections to having dogs on his property. I have never been denied access to land because I use terriers, and consider it more a matter of courtesy to explain with reasonable detail what I will be doing on his land. Without fail you will find that a farmer will appreciate being asked when and where you can take your terrier on his ground. Sometimes he may have restrictions concerning where he wants you to go, and it is always best to find this out before your day's hunting.

The best way to secure access to ground is through acquaintances; failing that, place a card in places that are frequented by farmers and landowners such as feed merchants, agricultural suppliers, rural shops and the local saddlery. Generally, if you become known to one farmer for doing a good job on his land, he will be able to put you in contact with many more farmers. It has been recommended in some books that you offer to help the farmer at busy times of the year, and in return he will allow you over his farm to rabbit during the winter. While it is admirable to be willing to help other people, I would not rely upon this, simply because most modern farms generally do not require extra manpower at busy times because they have the use of bigger and better machinery. Therefore, I would suggest that a direct request is the simplest and best way to obtain permission.

It can sometimes take a long time to gain access to good rabbiting land, and from experience I know how frustratingly dull it can be to be confined to second-rate ground where rabbits are scarce. When my brother and I first took up ferreting we had to trudge mile after mile just to get sight of a solitary rabbit, and seriously began to think that ferreting was the most boring interest in the world; people were understandably bewildered when they observed our jubilant celebrations when we actually managed to catch one. Nevertheless, the rabbiter should never succumb to the temptation to hunt on land without first securing permission, no matter how many rabbits are bouncing about upon it. Not only is it illegal, it will also ruin your chances of ever getting permission from anybody in the future and will incur the wrath of fellow field sportsmen who, in spite of acting responsibly, feel that their good reputation is in danger of being tarnished by this kind of behaviour.

I have lived in various different parts of the United Kingdom, and in some instances have had to wait over a year before I could secure really good ground that could be hunted regularly throughout the winter. I have also observed that, although sooner or later the rabbit population may not be seen in any number, invariably it is only a matter of time before new, abundant sites become available.

Alternatively, if you become frustrated with waiting for good hunting sites to materialize in your own locality, there is always the option nowadays of travelling some distance by car to an area where rabbits are known to be a nuisance. It is obviously more difficult to obtain permission in such situations due to the fact of not being known, but the concerns of suspicious landowners may be allayed if, for instance, a farmer from your own locality is willing to offer his recommendation and vouch for your conduct of the sport. You may not find this necessary, and may be greeted by farmers with open arms; but it is always worth having to hand when you have gone to the effort of travelling some distance.

Knowledge of the Land and the Indigenous Rabbits

Having gained permission to hunt over private land with your terrier for rabbits, it is always worthwhile getting to know its layout prior to the actual day's hunting whenever this is possible. In the long run this enables the rabbiter to save time and apply himself with more industry when it comes to hunting the rabbits. Having a good walk over the land will enable you to:

- locate the main areas of rabbit activity;
- identify rabbit warrens that are in use;
- select the best method in which to use the terrier: for example, sometimes it may be best to use him in conjunction with ferrets, at others with long nets or another terrier;
- take note of obstacles, excessive ground cover and any potential hazards.

It may even be worth making a rudimentary map to act as a guide. On this it would be easy to include information such as how extensive the warrens are, and therefore how many nets would be required if ferrets were to be used. The proximity of roads, barbed wire and electric fences could also be highlighted on the map. Among the sites where my terrier is regularly employed are a derelict farmhouse and a quarry, where local farmers have dumped old machinery, building materials and the like. It would have been irresponsible to put him to work at either of these locations without first making sure of what potential dangers there are and where exactly they are to be found.

A derelict steading with abandoned bales makes an ideal hideaway for rabbits, and a keen terrier is required to bolt them from their cover.

In order to track the rabbits down and make an accurate map, either on paper or in your head, depending how good your memory is, it is imperative that you have a basic understanding of the ways and habits of the rabbit. You should be able to identify signs of rabbit life, such as rabbit runs, diggings, seats and latrines, and know the most likely places for them to make their homes. Rabbits often prefer to remain above ground, but will still choose the protection of wood piles or long grass to keep themselves out of view. If this is the case on the ground where you hunt, you should pinpoint where these places of temporary refuge are so that you can flush them from their hiding places when it is appropriate. When a rabbit is made to bolt it will usually follow a trail or basic directional line to what it considers to be the nearest place of safety, and you should try to identify these trails so that you can either wait in ambush with your terrier, or attempt to drive it towards an area you are confident of being able to work.

Anybody who wants to hunt rabbits must make the effort to learn as much about them as possible. To start with, you would be advised to try a good book such as Bob Smithson's *Rabbiting*. Mr Smithson was a gamekeeper and teacher of environmental studies, and is of an age to remember rabbiting before the onset of myxomatosis. When he was nineteen he caught 10,600 rabbits in one year, and there is not any area of rabbit life that he is unfamiliar with; this is demonstrated in his helpful chapter regarding how to know your rabbit. It is also worth a look in second-hand bookshops because they usually have shelves full of inexpensive books on natural history; these are always interesting, and this is precisely how I have found some of my favourite reads.

If the intention is to use ferrets, you must be fully conversant with how to deploy them before you can turn your attention to the rabbiting terrier; and it goes without saying that you must have the skills and confidence actually to kill any rabbit you do catch. Traditionally you have the chop or the chin-up methods to choose from, and it doesn't really matter which, as long as you are efficient in how you go about it. It is not necessary for me to describe in detail the process of ferreting, or to offer point-by-point instructions on how to kill a rabbit, because there are a lot of books already available that have the time and space to deal thoroughly with these subjects, and they should be referred to if you require further commentary.

People who hunt animals on a regular basis will tell you that as a result of experience, and through observation, they have acquired a unique knowledge and respect for the animal they hunt. Although small, rabbits should never be underestimated. They can be cunning, resilient, tough and daring, and it is well worth knowing as much about them as possible before you attempt to pit your wits against them.

WHEN TO USE THE RABBITING TERRIER

Having equipped himself with a suitable terrier and secured land to hunt over, the novice rabbiter will want to know if there is a particular time in the year, month or even day when he is best advised to go hunting.

There is a tradition in Britain of confining the hunting of rabbits to the colder months because this is considered to be in the interests of the animals that are doing the hunting, these conditions being the best possible for work since they avoid energy-sapping hot weather. It also enables rabbits to have an uninterrupted breeding season, the higher temperatures in the summer months providing the young with their best opportunity to survive. Some research would indicate that rabbits are capable of breeding all year round, although this will depend upon how pleasant or severe the weather is. Hunters who confine their activities to the colder months will all know that, thankfully, they are rarely, if ever, faced with killing rabbits that are not fully mature.

It used to be said that you should only go hunting when the letter 'r' appears in the month. This gives a season lasting eight months, or sometimes a few weeks less depending on how warm September and April are. Some rabbiters like to wait until October before they start hunting, and base their decision upon the climate they experience in the locality where they reside. A late flourish or early onset of warm weather will result in a late start or an early finish to the rabbiting season. Nevertheless, the British climate always offers a season that provides us with ample time to reduce the rabbits to a number whereby the damage they cause is minimal, and I cannot think of any reason why this should change; therefore I only use my terrier for rabbiting during these winter months.

Within these months it does not really matter which day you go hunting on, or what the weather is doing on that day. The rabbiting terrier will work with enthusiasm when the weather is either frosty, windy, bitterly cold, pouring with rain or snowing, and during the course of a season most of us will encounter all the vagaries of the British weather.

It is pointless waiting for a day with so-called perfect weather conditions because you will probably miss several productive days' rabbiting while you are waiting; moreover, the dog is perfectly capable of adapting to the conditions in which it finds itself and would not thank you for delaying hunting out of consideration for its welfare. Inclement weather that would usually fill my terrier with resentment if I took him out for a walk, is completely ignored when his attention is turned towards hunting rabbits. Having hunted for some years, I have encountered fine days when I thought the weather would favour a good catch, and miserable days when I was convinced that I would not catch anything (except a cold), but in practice the reverse occurred, with the best catch happening on the worst day, and vice versa.

This shows that hunting animals with animals is not an exact science requiring a complicated formula to succeed. The rabbiting terrier is quite capable of, and should be given the opportunity to catch a rabbit in all weathers, although the conditions may limit the methods that can be employed – for instance, strong wind does not favour the use of nets in conjunction with the terrier. It should also be borne in mind that the weather will influence the strength of the scent that can be detected by the dog. Scent-laden air in contact with earth at a temperature lower than itself is cooled, sinks, and is gradually absorbed by the earth. All trace of the scent is then lost, and conditions for hunting are considered as unfavourable.

These situations can largely be overcome by the rabbiter possessing a thorough knowledge of the land over which he hunts and the most likely whereabouts of the rabbits. This will enable the handler to guide his terrier into close enough proximity to the rabbits to pick up the scent for itself, and when this is combined with the terrier's keen eyes and uninhibited inquisitiveness, the working terrier is able to locate and catch a rabbit in spite of poor scenting conditions. Scenting will be easier when the earth is warmer than the air with which it is in contact because this air is then heated and rises. The scent particles are not then absorbed by the earth, and conditions for hunting, which relies solely on scent, are then at their best.

What time of day to go rabbiting is chiefly a matter of personal preference and any restrictions imposed by the owner of the land on which you hunt. Obviously, if you are going to use terriers in combination with ferrets and nets, you will need a certain amount of good light. Dawn and dusk have proved productive times for a spell of opportunistic hunting with a couple of terriers, and a fair number of rabbits have been caught in just such a manner.

It is fortunate that the terrier is willing and able to work in all weathers because it provides the rabbiter with unlimited opportunities to hunt throughout the eight-month season. It is mainly a matter of the rabbiter's own preference, and what he is willing to endure himself, that will dictate when he does and does not go hunting.

EQUIPMENT FOR RABBITING WITH TERRIERS

Essentially, all that is needed to hunt rabbits with a terrier is the dog itself with a collar and lead. The lead is required when you have to cross or walk near roads or other potential hazards. Some terriers do not tolerate wearing collars and most prefer a harness for restraint, and this is the case with my own Jack Russell. If he wears a collar for a prolonged period of time, whatever it is made of, he will develop an irritation of the front side of his neck, the skin along the line of the collar becoming red and sore, and eventually resulting in hair loss if nothing is done to resolve the problem. This condition is peculiar to the fine- or smoother-haired breeds of terrier, but using a harness provides a simple solution, and my terrier can wear a harness all day and every day without any problems whatsoever.

When my terrier is working in what I consider to be a safe environment I always remove his harness, as I would also do a collar, so that he can negotiate all manner of obstacles without risk of getting caught up. I have seen enough Jack Russells follow rabbits into places where a collar or harness would quickly become caught, and when the dog is beyond the reach of the handler it becomes a potentially dangerous situation. A good terrier considers disappearing into unseen passages – whether through stacked bales, under woodpiles or into the thickest bushes – to be one of its primary functions. My crafty Jack Russell pursues rabbits underneath the floorboards in a derelict farmhouse where they think they are beyond reach; I was as surprised as the rabbits must have been when he disappeared into the tiny space between the earth and the floorboards, and this is a perfect example of why a terrier should, whenever possible, be worked without a collar.

Although the rabbiting terrier may disappear out of view for long periods of time, it will not go to ground in the same manner as its fox-hunting kin, and there is really no reason to consider the purchase of an expensive terrier locator, or to arm yourself with a vast array of digging equipment that you will very rarely use.

Apart from a collar or harness and lead, the only other piece of equipment I would consider to be an essential accompaniment to the rabbiting terrier is a first-aid kit for the dog, designed specifically to cater for the common injuries that can occur as a result of his industrious endeavours. Although there are tales from the distant past of tough terriers battling on against badgers despite losing an eye or having their jaw broken, there is no reason why a terrier today should continue any sporting activity without receiving appropriate treatment for any injuries that befall him. There are three reasons for treating injuries promptly; these are:

- for the comfort of the terrier;
- to enable the day's hunting to continue;
- to prevent further complications such as infection, which will delay the healing process.

Most terrier men that I have been acquainted with, including those who use their dogs for ratting, rabbiting and to hunt the fox, are quite fussy about their dogs and make every effort to prevent them getting hurt. I shall be looking at the items that should be included in the first-aid kit for the terrier, and also in a later chapter how to administer first aid, and shall therefore confine my comments at present to stating that it is an invaluable piece of equipment for anyone who goes hunting with a dog.

If the terrier is expected to do a full day's work, it should at some time have access to fresh water and a small amount of food in order to maintain its energy levels. I always carry a waterproof dog jacket to put on my terrier when working exposed sites in fierce weather conditions. I only do this because my Jack Russell is short haired and therefore amongst the most vulnerable of terriers to the effects of the winter weather. This is worsened by the fact that I live in an area that is notorious for its bad weather. When I lived in South Wales, the warmer weather meant that I never had to bother with such measures, despite it being incredibly wet for weeks at a time. A lot of terriers that live in the same part of Scotland as I do can be seen wearing jackets throughout the winter, and this is done perhaps because they are old, or because they are smooth haired. In all cases the terriers appreciate this protection from the elements, and there is nothing wrong with making your dog comfortable while it is working or during a break in the day for coffee or lunch. As we have seen in the previous chapter, there are numerous breeds of terrier whose coats are so weather resistant that they are in no need of supplementary protection; however, even these may like a quick rub with a towel after hours of exertion in the rain.

Some terrier men like to include a corkscrew dog tether, which can be secured into the ground and used as a fastening post for the terrier when you don't want it to wander off. My brother bought a corkscrew tether many years ago when we were new to the sport, and I do not think that he ever once used it, preferring to rely on his terrier's obedience to the command 'stay'. As a result, the tether is very much a tool of personal choice, rather than a necessity, and is probably of more use to the person with a novice terrier or with a number of dogs to be worked in rotation.

A lot of equipment that you may end up carrying will directly depend upon how you choose to work the terrier. Obviously, if you opt to use ferrets, you will be loaded with a carrying box and bag full of nets, but even in such cases you should resist the temptation to overburden yourself and should take only what you need. Since we hate digging, my brother and I rarely take any spades with us, and only as many ferrets and nets as are absolutely necessary. The person who is using a number of terriers and is on his own will have no need to carry additional equipment.

A GUIDE TO USING THE RABBITING TERRIER

This is probably one of the least well discussed areas in other books that consider the topic of working terriers. When it comes to describing what to expect from the rabbiting terrier and the role that the handler should play, the information is undeniably vague. Perhaps this is because rabbiting is an uncomplicated sport that has been mastered in the past by children and adults alike, and a bit of thought and common sense go a long way to achieving success. Without doubt, practice and experience are the best teachers of any sport, but some guidelines and advice always proves helpful. The following information is intended to answer two fundamental questions:

1 What should my terrier be doing when rabbiting?
2 What instruction should the handler be giving the terrier whilst it is working?

It will probably be easiest to do this by discussing the roles of the rabbiting terrier separately, and I shall begin by discussing its most typical use.

Working the Terrier in Conjunction with Ferrets and Nets

Ferreting is concerned with the warren or home of the rabbit, so when used with ferrets, the terrier's sphere of work will fall within the area immediately around a warren and the approaches to it. It is always best to approach a warren from a reasonable distance away if possible, because this will prompt any outlying rabbits to bolt towards the warren, rather than away from it. Imagine a typical medium-sized warren nestling in the corner of a 4-hectare (10-acre) field that provides grazing for sheep – the field has one tiered hedge leading to the warren, and the other three sides are fenced.

The terrier should be free to work the outer margins of the field as it moves towards the warren. Its task is to identify anywhere that rabbits might be hiding, and to make them bolt to the warren. It is surprising how rabbits can hide themselves quite easily from the human eye by simply nestling down into tufts of long grass that they fashion around themselves like a canopy. These are known as squats or seats, and on several occasions I have nearly stood on top of a rabbit because it is so well hidden in one of these grass sanctuaries – and if it can escape attention by remaining motionless, it will not move, however close a person may be to it.

The rabbiting terrier has a natural inclination and desire to search out these hiding

The working terrier checking the side of a field where rabbits sometimes make themselves a 'seat' in the long grass.

places, and will of its own accord work methodically along a hedgerow or through gorse to see if anything is there. Do not imagine for a moment that a field like the one described will only have rabbits concealed in the hedgerow: the sides of the field that are fenced may not offer the same amount of cover, but the grass does tend to grow long at these edges, and I have lost count of the number of rabbits that my dog has bolted from squats that are right next to a fence.

When negotiating an expanse of open land, the terrier may make you almost dizzy as it criss-crosses the ground in search of rabbits; it will cover an enormous distance as it does this, with the purpose of flushing out a rabbit or fastening on to a strong scent of rabbits. There is usually no need to interfere with the terrier while it is doing this, and I do not even bother disturbing my terrier's attention by offering words of praise or encouragement: these would clearly be superfluous, because the terrier is happy with what it is doing and has no inclination to stop.

The main role of the handler during his terrier's search of the field is to give his dog the time to do its job properly. Being a successful rabbiter is not about speed, and it is absolutely pointless to sprint across the ground like an Olympic athlete, incessantly calling your dog to follow closely at your footsteps. The terrier must always be allowed the time to do its job in a controlled fashion.

In the unlikely event that the terrier refuses to make a search of either hedgerow, fences or coverts, and heads directly for the warren, the handler should intervene. It is well worth slipping a piece of cord through the terrier's collar and walking it along the edges of the field until finally arriving at the warren. I use a length of spun nylon threaded through the collar with both ends held in one hand so that the dog can be immediately released should it come across any rabbits. I have used this method on numerous occasions, and the speed of release enables the terrier to move as it wishes without delay. Although cheap baler twine will prove perfectly adequate for this job, I would advise you to select a material that will not cut into your hand. By taking the terrier round the field in this way, guiding him with a simple string lead, it will soon learn what is required.

Even if no rabbits bolt, the terrier must always begin its work with this search of the

The rabbiting terrier using its nose to follow the trail left by a rabbit.

outlying area. In the event that a rabbit is bolted, either the handler should have made sure that it is perfectly safe for the terrier to chase the rabbit towards the warren, or he must be totally confident in his ability to recall his dog. It is no good having a rabbit chased down a road by an over-excited terrier, which in turn is being chased by a man shouting desperately for his dog to come back. Control of the dog without undue interference must be the terrier man's guiding principle, at all times and in all circumstances.

Having efficiently searched out the land on the way to the warren, the terrier can turn its attention to the warren itself. It is expected to mark a warren that is in use even when it does not see rabbits bolting into it, and it does this by means of its ability to scent. By 'marking', the terrier indicates by its posture, demeanour and enthusiasm whether there are rabbits lurking in the underground labyrinth or tunnels, and it does this by sniffing in the air at the mouth of each bolt hole.

The terrier must refrain from immersing itself up to its hindquarters in a bolt hole, and from entering the warren itself, like a ferret, when the entrances are large enough. When there is no strong scent of rabbit, the terrier will display an obvious disinterest in working the warren. It has been known for rabbiters to ignore their dogs, but the terrier always proves to be right and is quickly relied upon as a precise guide regarding whether the warren is or is not worth working; and this is the fundamental purpose of the mark.

There are distinctive signs to look for when the terrier is marking an active warren. These are as follows:

- a prolonged period of smelling the air at the holes of the warren with the terrier seemingly drinking in the scent;
- the terrier standing at the bolt holes with its head pointing down them attentively. Movement of the head from side to side might be observed as the terrier attempts to hear more clearly;
- an idiosyncratic wag of the tail. This is not the same frantic wag that the terrier offers as an excited welcome, but a systematic rhythmic movement like the pendulum of a grandfather clock;
- sometimes my terrier uses his voice – when he can detect rabbits he grumbles constantly, and the closer he is to a rabbit the more intense the grumbling becomes. I do

A traditional rabbit warren suitable for using ferrets. Here the terrier should 'mark' to signify if the warren is in use, and be in a position to catch any rabbits that might push a net aside.

not discourage this because it is an exact and easily understood method of communication between dog and handler.

In order to mark effectively, the terrier must be allowed free movement around the warren, and the main benefit of this will be the use of its far superior powers to detect well hidden bolt holes that, without its efforts, would have been missed and therefore left without a net to cover them. Once again the terrier requires no direction and little praise when doing this, because it is a task that the dog believes itself born to perform, and it derives immense enjoyment from being given the opportunity to do it.

If a novice terrier is prone to over-excitement, or runs over the warren haphazardly without marking it properly, the handler has two options available to remedy this. Firstly, he can secure his dog to a leash and take it slowly and methodically around a warren that is known to be inhabited, and encourage it to sniff at the holes. After repeating this a number of times the terrier will calm down and can be trusted to conduct himself correctly off the lead. The other option is to work the dog alongside an older and more experienced terrier that will provide the novice with something to imitate. Appropriately, the guiding terrier is referred to as a 'schoolmaster'.

Some sportsmen are concerned that the movement of the dog over the warren will alert the rabbits to the presence of hunters above ground, which will therefore make them reluctant to bolt; however, from my own experience I would suggest that these worries are unwarranted. It should be considered that, following the thorough and accurate marking of a warren by a terrier, the dog will be withdrawn while the purse nets are being set, and this provides time for the rabbits' fears to subside, if indeed they have been aroused by the muzzle of a dog appearing at all the bolt holes. I have hunted in situations where the rabbits would have to have been blind, deaf and stupid to remain unaware of the posse of men and dogs marauding about their home – and yet they still bolted, thanks to the perseverance of the ferrets. I abide by the recommendation that a minimum of noise and disruption is best when preparing to work a warren, but do not believe that the activities of a well-trained terrier fall within either of these categories.

The best teacher of a terrier is often a more experienced terrier.

As already mentioned, the terrier must be made to sit out of the way so that a purse net can be set over each and every hole. If he is allowed to continue sticking his nose into the bolt holes he will make a thorough nuisance of himself and will ruin all efforts to set the nets – and you will encounter few aspects of hunting that are more infuriating than the constant disruption of nets you have carefully positioned, particularly when it is your dog

The terrier must remain on the alert, but be out of the way when placing purse nets.

that has caused the trouble. It is therefore the handler's responsibility to let the terrier know that it should remain clear of the nets while they are being set. I do in fact allow my Jack Russell to go where he wants while I am placing the nets because he has the knack of being able to walk more or less completely over a net without disturbing it, and will keep his nose on the outside of the net.

With all the nets in place and the ferrets entered, the terrier should be allowed to move from the 'sit' position so that he can station himself where he can guard and watch the bolt holes. He will be able to sense when and where a rabbit is going to bolt from some moments before it actually does, and will indicate this by his expression and posture. I only had to observe my terrier once or twice

The terrier must be taught to sit and watch a bolt hole so that he can be used where you cannot place a net.

before I realized what this idiosyncratic behaviour signified. It is similar to the phenomenon of knowing what a close friend or relative is thinking by the look displayed on his or her face. The experienced owner is able to interpret the particular expressions and stances adopted by his terrier as clearly as he can read a book or hear a voice. By keeping one eye on the dog and the other eye on the warren, I am able to get myself in the best position to deal with a bolting rabbit.

The terrier must possess the self-restraint not to charge at the rabbit when it is just about to bolt or when it is caught in the net. If allowed to do so when it is about to bolt, the rabbit may well turn back into the depths of the warren and will then be more difficult for the ferrets to move. The terrier should not be permitted to grab a rabbit that is entangled in a purse net because it will get in the way of the ferreter who, in this situation, is able to kill the rabbit more quickly and without tell-tale terrier damage being inflicted on the body. I keep control of my terrier at this stage of the proceedings by making sure that he remains just fractionally behind me, using the command 'stay' to reinforce this, and to stop him pre-empting my actions. By so doing, he is kept close enough to burst into action and deal with the rabbit in the unlikely event that the net should fail to purse around it.

This is the terrier's function when working a basic rabbit warren in conjunction with ferrets. It may appear to be a minimal contribution, but its seemingly simple task increases both the chance of catching rabbits and the number of rabbits that can be caught. Only those who have seen a rabbiting terrier doing its job will appreciate the multitude of little things that the dog does so well, thereby earning it the admiration of all terrier enthusiasts. It goes without saying that the terrier and ferret should have previously spent enough time in close association to develop a relationship of mutual toleration, which enables them to act as allies with a common task when hunting, so there is no real need for the handler to give his terrier constant commands to stop it attempting to devour the ferret.

Having spent over two decades using ferrets to catch rabbits in a variety of different environments, I know that all warrens are not equally easy to work, and at times my brother and I have been left scratching our heads wondering how in the world we are

The terrier's ability to make its way through stock-proof fencing is essential.

going to work some of the warrens that we were faced with. It is in such circumstances that the terrier proves to be invaluable, as it performs tasks that no other dog could.

When ferreters who do not have a dog are confronted with bolt holes that they cannot reach, they must either leave these without nets and run the risk of having the rabbits escape by this exit, or not bother working the warren at all. On all of the ground where I have used ferrets, there have always been some bolt holes under gorse or amongst abandoned farm rubbish that it would take a month of clearance to get at. However, I have always managed to work these warrens thoroughly due to the versatility of the well-trained rabbiting terrier. This terrier is used like an extra net, and takes up a position next to the bolt hole from where he will be able to snatch any rabbit that shows itself. It is for these situations that the terrier is taught the command 'stay'.

Terriers like these are in possession of a natural cunning and full of working instinct, which is what makes them so useful.

Although some rabbits bolt at a tremendous speed, a waiting dog will always know they are coming because of the thumping noise of their hind legs against the narrow tunnels of the warren. Thanks to the acuteness of the terrier's hearing, and the unrivalled dynamic speed with which they can move in confined spaces, they will prove to be as reliable a catcher as any net. You might think that the rabbit will sense the waiting terrier and turn back into the depths of the warren, refusing to bolt. From my experience I have found this not to be the case, which I would suggest is for the following reasons. The terrier is told to stay in a position that does not overshadow the bolt hole, but is close enough to be able to catch the rabbit. An experienced terrier has cunning and seems to know the best time to make its move. Unlike other dogs, such as my brother's collies, which will recklessly lunge down a bolt hole the moment they hear anything of a rabbit, the terrier has the sense to realize that he is much more likely to catch a rabbit just as it is leaving the warren. Even if a rabbit does try to turn round in an effort to get back into the depths of the warren, the terrier has the speed of reaction, which enables it to thrust itself down the bolt hole and grab the rabbit.

Obviously, it you are unable to reach the hole to place a net on it, you will not be able to get hold of the rabbit to dispatch it, and this is one of those situations when the terrier should be allowed to kill the rabbit. I remember when my Jack Russell caught a rabbit under some sheets of plyboard dumped in a quarry, my brother and I began to feverishly throw the rubbish out of the way thinking that it was our job to kill the rabbit. It was a monumental waste of time and effort that we will not repeat, because the terrier only took a few moments to kill it.

It is equally pointless to try and train a terrier to retrieve a live rabbit to you. It goes against all the laws of their nature, and the initial bite with which the terrier has seized the rabbit will have done the major damage anyway. I believe it to be much quicker and more humane to let the terrier kill the rabbit without interference. Once the rabbit is dead, the terrier should either be told to leave it so that it can resume its duties, or be recalled to the handler. There is really no need to teach the terrier to retrieve a dead rabbit, because there is absolutely no way that he will leave it behind, especially when he will be so proud of his work. My terrier has been known to pull rabbits that the ferrets have killed out of the warrens and wood piles without any prompting or guidance. It can be difficult to get a terrier to release a dead rabbit, however, and this is when earlier training with food removal (*see* p. 113) proves its worth. During his early outings my terrier would abscond with the rabbit he had caught when he thought no one was looking. If you are going to work your terrier in the manner described, it goes without saying that it must be 100 per cent trustworthy with ferrets because a ferret may appear at the bolt hole and should not be mistaken for a rabbit.

When your terrier becomes experienced at rabbit hunting, he will develop his own way of doing things and seems to know instinctively what it is best to do in each situation. When the terrier has attained this level of competence, the handler should have the confidence to let it dictate its own actions. Only when you are willing to do this will you see moments of pure terrier genius at work.

My brother's old Jack Russell used to be allowed to patrol the warren as she wished, and she was often able to assist the ferrets when rabbits were proving elusive. She had an uncanny knack of being able to dart into the warren and snatch a rabbit as it moved near the surface. However, she would never dream of grabbing hold of a rabbit caught in a purse net, and my Jack Russell also has the manners to leave rabbits that are netted to me, although he will be watching very closely just in case anything goes wrong. Both these terriers appear to be able to walk over nets without ever disturbing them, and I have noticed that the more you trust them to get on with their work without constant interference, the better behaved they are.

WORKING A TERRIER ALONG A HEDGEROW

At one time rural Britain was criss-crossed with a myriad of hedges; however, with changes in agricultural practice, a great many hedges were uprooted as fields became enlarged to accommodate large agricultural machinery. During the past decade, there has been a revival of interest in hedge planting and laying, and those hedges that date back many decades will be left where they are.

I have always been fond of traditional hedges, and I am joined in my appreciation by rabbits, which will either make their homes amongst the roots of the hedge, or use the cover it provides as a temporary shelter or seat during the day. They seem to like the thickest and most neglected hedges best; many years ago when I lived in Cambridgeshire, blackthorn and hawthorn were the most common hedging materials that I encountered, whereas in the north-east of Scotland, gorse and broom are more frequently seen. On occasion I have been obliged to force my way through a thick blackthorn hedge, and without the protection of thorn-proof clothing it is not a pleasant experience or one that I intend to repeat. This is hardly surprising when you consider that the original purpose of the hedge was to contain livestock of a much greater size and thicker skin than myself.

It follows that a dog that is expected to pursue a rabbit through what can aptly be described as nature's answer to barbed wire, must be tough enough to ignore the prickly briers as they jab and scratch the body. When it comes to toughness I believe terriers are in a league of their own, despite the fact that they are not the biggest, strongest or most powerful of dogs. In truth, making headway through a hedge suits the smaller breeds more than the larger ones, so working a hedge in pursuit of rabbits is generally considered to be the task of the working terrier and is therefore left to him.

Surprisingly the terrier relishes the task, and I have read of a pair of Irish Terriers who frequently absconded from their garden to hunt, under their own initiative, up and down a nearby hedgerow, and they did this with a fair degree of success. But under supervision, what exactly does the handler expect of his terrier? Essentially his task is to shift the rabbits to one of two destinations, depending upon which hunting method is being employed: if ferrets are to be used, the terrier should be flushing rabbits from the cover of the hedge and driving them towards the nearest warrens. Obviously, in order to do this properly, the handler should ensure that his terrier begins its task at the point of the hedge that is farthest away from the warrens, to which it is hoped they will be driven.

The second method of hedgerow hunting involves using another dog. The terrier will be used once again to move along the hedgerow seeking out rabbits, but this time he will be travelling in the direction away from the warren. Another dog with more pace, such as a whippet or one of the faster terriers, will be following the path of the terrier from a distance. The intention is that the terrier will flush the rabbit into the path of the waiting dog. A thorough knowledge of the ground you are hunting will provide you with the best indication of where to position the waiting dog. Rabbits when bolted do not just run haphazardly, but tend to follow particular, well used trails. Being able to identify these pathways will give you the ideal place to locate the dog so that, with a short burst of speed, it will stand a good chance of catching the rabbit in the open.

Inevitably, the terrier that is hunting a length of hedge will lose sight of the rabbit that it bolts, and the more far-sighted handler must observe the progress of the fleeing rabbit, so that he can redirect the terrier or deploy his ferrets, depending upon the rabbit's final destination.

A terrier working a really thick hedge may bolt a rabbit for a waiting gun if there is sufficient open ground. Often the terrier will continue to search the hedge while a rabbit is

A terrier at work, making its way through a hedge.

making a hasty retreat, and therefore an adequate distance will have opened up to allow a safe shot. The obvious danger is that, unlike with birds, you will not be shooting above the dog, but in line with it. Traditionally white terriers were preferred when working with a gun for ease of identification, and certain breeds of terrier are more suited to the gun than others. The Irish terrier established a reputation as the best type to accompany a shooter.

Increasing the number of terriers will improve the chances of actually catching a rabbit within the hedge, because they can hunt both sides at the same time, and thereby reduce the rabbits' chances of bolting. Some rabbit-hunting men have been known to combine a dog of hound blood, such as a Beagle, Basset or Dachshund, with a couple of terriers when working a hedge. The logic of this type of partnership is that the hound with its greater tracking ability will be able to follow the trail of rabbits when they have moved outside and evaded the terriers' radar.

A lot of field sportsmen enjoy watching their dogs hunting hedgerows because of the demands it places upon them. It is an ideal opportunity to watch the terrier use all its energy, ability and skill to catch a rabbit.

WORKING A TERRIER IN DERELICT BUILDINGS AND FARM RUBBISH

I was fortunate enough to have been given the rabbiting rights to farmers' land when I lived in England and Wales, and now that I live in Scotland, I can say that farmers up and down the country typically allocate a place where they lay up their old and disused machinery. Vegetation grows up and envelops the machinery, providing a veritable haven where rabbits can dig their homes out of the easy reach of man. The farmer may also store an assortment of objects that he intends to use in the distant future such as railway sleepers, wooden pallets, corrugated tin roofs and old fence posts. Sometimes these are abandoned without any method, and are adapted by the rabbits as building materials.

At one of the sites where I hunt rabbits is a derelict farmhouse, its neighbouring barns full of old and broken up bales of hay, and a small quarry where rubbish has been dumped for some time. Ironically, the rabbits are at their most dense in amongst all this debris – and flushing them out of their

An old quarry colonized by rabbits, presents specific challenges to the rabbiting terrier.

stronghold requires a slightly more innovative approach than does hunting the traditional rabbit warren.

In such a location the main role of the handler is to ensure that it is safe for his dog to work, and that he can retrieve the terrier if needs be from the many hazards that are present. There is inevitably broken glass within derelict houses and disused farm buildings, and often discarded barbed wire and wood with rusty nails protruding out of it – so take precautions that the terrier does not injure himself on these materials. You should also take note of any abandoned drums that may have contained chemicals in the past, and make sure your terrier stays well clear of them. Basically, if you have any doubts about safety, the advice is simple and emphatic: your terrier's health is not worth putting at risk, however many rabbits you think you may be able to catch.

A farmer drew our attention to the fact that a large number of rabbits was living around the derelict farmhouse, and spreading. When we walked around it during the day, the number of rabbits that bolted proved the point, and we were expecting to find a massive warren that they were using as a refuge on the outlying land. But all we found was a couple of small warrens with the occasional rabbit in. Being of a nosy disposition and also interested in old buildings, we went for a look inside the derelict house. Inside what would have been the sitting room, my brother leaned over to move a piece of wood off the floor, and as he did so, a rabbit leapt in the air and disappeared through a hole in the floorboards.

It was obvious that the ferrets would be no use in such an environment: they would end up chasing the rabbit round and round in such a large space, and we would have no idea where they were, short of ripping the whole floor up, and no way of getting them back. So we let the terrier have a look, to see if he could do anything – and without a second's thought he disappeared under the floor where there was about a 20cm (8in) gap between the floor joists and the earth. The lack of space did not deter him, and we could hear him moving around underneath the floor. He drove the rabbit into a corner and made us aware of where he was by barking and whining; by his constant grumbling and baying we knew he was on to a rabbit, but obviously did not have sufficient room to kill it.

This derelict farmhouse may not be the obvious place to find rabbits but they are to be found in large numbers under the floorboards.

We cautiously removed a little piece of skirting board and prised the end of the floorboard away from the underlying joists: a rabbit stood upon a foundation stone that acted like a ledge. My terrier could not quite reach this ledge, but he did have the rabbit well and truly trapped. My brother reached in and extricated the rabbit, just like a magician pulls one out of the hat. The terrier's grumbling did not cease, and, as a result, the arm went under the board again and withdrew another rabbit – three were caught in the space of as many minutes in this way. When the terrier finally stopped grumbling, we thought we would have to pull up the floorboards to get him back; but he had remembered where he went in under the floor, and quickly returned to us.

This type of work is typical of the terrier's usefulness. When you think that the rabbits have got the better of you, the terrier will do something to change this, usually on its own initiative. Invariably in such situations there is nothing you can do to help the terrier, and your energies should be directed to ensuring that the environment is safe for him to work in. You should also make sure that you can detect his progress, if not by sight, then by sound. It was only by listening carefully that I could deduce what my terrier was doing under the floorboards: I could clearly follow his progress from where his body rubbed against the underside of the floor, and could therefore follow him, nearly step by step. My Jack Russell uses his voice quite naturally and without any encouragement. If he has a rabbit cornered in a confined space, he will grumble constantly until it is caught. I must confess to liking my terrier's use of his voice because it is as good as a person telling me in plain words what is going on when it is out of view.

When hunting rabbits that take refuge in this derelict house and dilapidated barns, we have no option but to stand back and let the terrier get on with it. Terriers seem to have an intuitive knowledge of when we are relying on them, and always rise to the challenge. However, there are two things that are worth bearing in mind: firstly, if you start to lift any floorboards or slabs, make sure they cannot snap or fall and injure the terrier; and secondly, working in these cramped conditions is extremely hard work for your dog, and you should avoid overtiring it.

Catching rabbits amidst an assortment of farm rubbish is another activity in which the

Searching out farm rubbish where rabbits are often to be found.

working terrier proves its ingenuity and shows its capacity to adapt to unusual hunting environments. The most common farm waste that I have encountered includes archaic bits of machinery, sheets of corrugated metal, old wooden pallets and piles of boulders – and increasingly, I am finding that rabbits are using these materials as cover for their warrens. As a result, purse nets and ferrets can only do so much because it is virtually impossible to identify all the bolt holes in a mound of rocks, and those holes that are obvious are, more often than not, very awkward to cover over properly with a net.

My brother and I have achieved a fair amount of success by continuing to use purse nets in these locations, even though we have often had to improvise by placing a rock on the net peg, or by sliding the peg between two sheets of metal in order to secure it. With all the obvious exits covered, we enter the ferret to see if it can shift the rabbits. In these circumstances the terrier is encouraged to stand unusually close to the exit hole because, unlike simple warrens where pegs are driven into the earth, there is a reasonable chance of net failure. In this event, the terrier will either be able to snatch the rabbits as they push the nets aside, or, by blocking their path, will force the rabbits to turn around, and in their panic, as they attempt to re-enter the warren, they will entangle themselves in the nets; this process is commonly known as back-netting.

If the terrier is to be useful at these sites, it must be allowed the freedom to move around as it sees fit. Consequently, both terrier and ferret must be on the best of terms because there is a good chance of them coming nose to nose with one another. The main advantage in the terrier's favour is the noise the rabbit makes when it moves through rocks or under sheets of metal. The terrier's sensitive hearing can accurately locate the source of such noises, and it can then wait in ambush for the rabbit.

We have hunted rabbits in huge mounds of rocks and have noticed that they will try and hide in small pockets in the rocks, and will remain absolutely still in an attempt to escape detection. My Jack Russell will locate them by using his nose to good effect, and by his body language he will clearly signify his find. Often all we have to do is move a rock slightly, reach a hand in and extract the rabbit. Sometimes the ferret will kill a rabbit

A warren located in the remains of an old stone wall is impossible to net precisely, and the terrier must be waiting nearby to deter or drive back into the net those that would otherwise escape.

that refuses to bolt from its cover, and again my terrier will use his nose to good effect by retrieving dead rabbits from wood piles and from under tin.

There are no clear rules as regards working at such locations, and sometimes the rabbiter only learns the best way to direct his team of ferrets and terriers by trial and error. This has been our experience. It is, however, certain that if you trust your terrier, and give him his freedom to use his talents, he will reward you time and again.

WORKING THE TERRIER WITH OTHER DOGS

The primary function of the rabbiting terrier does not radically alter whether it is working with ferrets, other dogs or as part of a pack. The terrier's essential job is to locate the rabbits and either catch them or bolt them to a location where they can be caught. We have seen that the terrier is an invaluable member of the ferreting team and helps to influence the outcome of a day's sport. Now we have to ask ourselves if there is any point to partnering the terrier with a dog or dogs of another breed.

From experience I would say that a dog of virtually any description would prove to be a help to the terrier when working with or without ferrets. I used to work my old mongrel alongside my brother's terrier, and now his collies regularly accompany my Jack Russell when we go rabbiting. Such dogs may have their limitations, but even so they lessen the workload on the terrier and in some cases ensure that its work is not in vain. For instance, my Jack Russell will often track a rabbit down and force it into the corner of a barn where there are fallen roof tiles and other debris that it can hide under. There is always more than one way out of these temporary refuges for the rabbit and no obvious places to put nets, and because the terrier can only enter from one side, an assistant who can guard the other side is just what is needed. He will be able to prevent the rabbit bolting, which gives the terrier time to make its way into the debris and seize hold of the rabbit.

Using a dog with extra pace together with a terrier will obviously give the sportsman more opportunity to catch rabbits, and it should not be forgotten that the lurcher was always considered to be the archetypal pot

It seems that all dogs, even my mother's laid-back spaniel, are excited by hunting rabbits.

filler. The terrier's task is to do the so-called 'undercover' work and penetrate those obstacles, natural and otherwise, where rabbits hide in order to force them out for the waiting dog, and the terrier is the dog of choice for this type of work. It is the terrier's natural inclination to explore these hiding places, and I do not have to offer my Jack Russell any encouragement or instruction when he is working these locations.

The dog that is being worked with the terrier should be held back by voice command or a slip lead in a position where it will be able to see the rabbit bolt and thereby offer chase. If the dog starts to move before you see anything, do not stop him, because his anticipation will be based on what he can hear, which is far more than we can, and the head start this gives him can mean the difference between catching and losing a rabbit. However, you should not allow this dog to immerse itself in the ground cover in an attempt to do the same job as the terrier because it will not be able to do as good a job, and it is not what it is there for.

Now we come to the tricky question of which breed should be allowed to work with the terrier. As I mentioned earlier in the book, I have opted to abide by the wisdom of Brian Plummer and avoid working longdogs and terriers together. It should be noted, however, that there are others who say that if you bring these two up together so they have forged a good relationship, they will make a good working combination without being fiercely competitive.

Of the other breeds of dog, I would not disqualify any of them from going rabbiting. Every dog seems to love hunting rabbits and, given the opportunity of regular outings, will soon learn the tricks of the trade with the terrier acting as mentor. I will not work any dog with my terrier unless they are firm friends and willing to co-operate with one another, and I would advise you to do the same. Even if someone claims to have the best rabbiting dog in the county, if it does not get on with your terrier, do not on any account go hunting with it.

I would suggest that the best breeds to be worked with terriers are the Whippet, the Border Collie, the Basset Hound, the Beagle and the Dachshund:

- the Whippet, because it offers pace and is synonomous with rabbit catching;
- the Border Collie because it is very obedient, quick to learn, has remarkable stamina, and is good with livestock;

This terrier is a very hard worker, but by herself, there is only so much she can do.

The collies' strengths are their obedience and ability to endure the weather.

- the Basset, Beagle and Dachshund because they are instinctive rabbit hunters with a fantastic ability to scent and track rabbits; however, they can be difficult to control.

I have heard about terriers becoming embroiled in fights when they are made to work with other dogs, and sometimes they have been the aggressor and sometimes the victim. But despite what some people may think, the working terrier does not have an appetite for conflict, and is much more inclined to co-operate with other dogs so that the objective of catching rabbits can be achieved. My Jack Russell puts all his energy into his work and does not let his attention be diverted to waste his time on what he would probably consider to be petty issues. The collies that work with him recognize his sphere of work and have slotted in where they feel they are at their best – and there has not yet been any angry exchange during this, their first year of working together.

These collies have proved to be useful companions to my terrier, and have managed to account for some rabbits themselves.

For those who wonder if the collies have proved to be of any use, I can state that they have caught a small number of rabbits themselves, they work well with ferrets, and they complement the terrier who really taught them what it is all about.

WORKING A PACK OF TERRIERS

A rabbiting terrier is an industrious and unrelenting worker, but there are limits to what one terrier can achieve. Having two or more terriers will increase your chances of catching rabbits and reduce the workload previously borne by a single terrier, so that you can hunt for longer periods of time. Although another breed of dog may prove helpful to the terrier when rabbiting, only another terrier will be able to share in every aspect of its work, and this is a direct result of the size, manoeuvrability and instinctive hunting behaviour of terriers.

My Jack Russell can squeeze himself into some places where all that my brother's collies can do is stand by and cheer him on; try as they will, they just cannot follow him because they are too big. Furthermore, they cannot respond to a fleeing rabbit or alter their direction of travel as quickly as a terrier. I have noticed this time and again during games where the big collie dog chases my Russell: speed is not a problem for the sheepdog, but just as he closes in, the terrier will jink right and left like a hare, and the collie will either over-run or tumble over as he attempts to make the turn.

If you are going to rely predominantly on terrier work to catch rabbits, then you should aim for a minimum of two. They will be able to approach warrens and work hedgerows and obstacles where rabbits hide, from opposing directions and thereby cut off the escape routes. As generations of terrier owners have discovered, you really do not have to teach terriers the correct way to hunt together, because they seem to realize the best way that they can assist one another, and terriers, particularly when they are of the same breed, exhibit a pattern of behaviour when hunting that would make you swear that they had discussed tactics beforehand. I have also noticed that terriers work at a common rate, especially along field edges and hedgerows, as they endeavour to be as thorough as possible, literally leaving no stone unturned. This is in

Only the terrier has the balance and size to work these kinds of rocky outcrops.

contrast to my brother's collies that cover the ground like a sprinter, and need to be reminded to hold themselves back.

If your intention is to use the terriers alongside ferrets, I would suggest that two to four terriers would be an ideal number. This is not so many that they would be unmanageable when you are deploying ferrets in a warren, and yet they are enough to be able to hunt effectively in locations where ferrets cannot be used, such as derelict houses, amongst farm waste and in wood piles. Four terriers are also manageable in terms of expense and day-to-day care.

When choosing terriers to form a small team, you have the option of either selecting from the same breed of terrier, or mixing different breeds of terrier. The advantage of the former is that terriers of the same breed seem to have an instinctive mutual understanding and take to each other quite readily, probably becoming friends more quickly than they would with different breeds of terrier. A pack consisting of the same breed could be acquired from the same line and maybe even the same litter, in which case they would be reared together.

On the other hand, there are some people who maintain that when you select your pack from only one breed you are limiting yourself to the talents peculiar to that breed, and that there is greater advantage in putting together a team of different breeds. Not only can you combine each one's inherent hunting talents, you can put together terriers of different sizes and speeds and thereby compensate for the weaknesses of one breed with the strengths of another. For example, you could select a Jack Russell or Norwich because they are excellent all-rounders, a Dandie Dinmont for its superior ability to scent, an Irish or Bedlington for its speed, and a Border to hold the team together, or maybe a Yorkshire Terrier because of its diminutive size and ability to get into the places that even other terriers can't. There are some sportsmen who do not like mixing different breeds of terrier together in one pack, but they cannot deny that such a band of assorted terriers can be very effective when hunting rabbits together.

Terrier men do seem to have their favourite breed or breeds of working dog, and trying to make them alter their opinion is like trying to move the Rock of Gibraltar! I have known of some hunt terrier men who like small, lightweight terriers on short legs, and others who prefer more substance or length of leg. There is absolutely nothing wrong with just wanting a type or breed of terrier that complies with your view of what a perfect terrier should be, and this is why some people stick to a team that is made up of only one breed.

I do not really see the point of having a pack of more than six terriers for rabbiting nowadays, for a variety of reasons. First of all, hunting with a pack is not a particularly

A mixed pack which will offer a greater variety of skills.

effective way of controlling rabbits, and the number of rabbits that you may catch in this fashion will not be enough to satisfy most landowners. It is also worth bearing in mind that large packs, numbering well into double figures, were used in the pre-myxomatosis era when the widespread distribution of rabbits throughout the United Kingdom was markedly different to what it is now. These large packs were also used for other sporting activities and it is doubtful whether many contemporary field sports' enthusiasts will have the time to give these packs the hunting opportunities they deserve.

Whether hunting rabbits with two, four, six, twelve or more terriers, the same fundamental rules apply, and these are as follows:

- Never venture out with a pack, no matter how small or large, that you cannot control.
- Only combine terriers in a pack if they have a good relationship with one another; avoid using terriers that are bullies and any that behave unreliably, rushing around causing fights and delaying the hunt.
- Use the terriers wisely in locations where they have a chance of catching rabbits, such as hedgerows and other sites where rabbits tend to use the ground cover instead of warrens.
- Work a pack starting at the closest point to the warren that you can get, and work in a direction away from the warren in an effort to catch the outlying rabbits and prevent them going to ground.
- Allow the pack to work naturally – even if you have one breed of terrier, you will find that the members of the same team have different strengths. A leader will probably emerge and this should be encouraged.
- Do not let one member of the pack interfere with other members of the pack; if it does, you must redirect its energies, preferably with voice control.
- The pack must be kept under control at all times. Each member must be thoroughly trained and, as a group, they must maintain their discipline. If there are any doubts that this can be achieved, they should not be taken out together.
- You should always keep the pack in view.
- You should have confidence in the working abilities of the terriers and trust them to get on with hunting the rabbits.

I think a pack of between four to six terriers is ideal for the modern rabbiter.

CHAPTER FIVE

Training the Rabbiting Terrier

THE OBJECTIVE AND PHILOSOPHY OF TRAINING

This chapter is based on a series of articles that my brother wrote first for the magazine *Gun Mart* and then in more detail for *Countryman's Weekly*. The credit for what is written therefore belongs to him – although in typically brotherly fashion, I could not resist making the occasional addition. It may be helpful to the reader if I briefly explain how my brother and I developed our training programme for our rabbiting terriers. First we observed the behaviour of terriers that were used to hunt rabbits, paying special attention to the praise and reproach bestowed on the terriers by their handler. It was as interesting for us to see what a terrier will do that gets it into trouble when hunting as it was to witness those moments of instinctive genius that fill the onlooker with admiration for the industrious little chap. Secondly, we took note of our own experience of rabbiting and decided how we thought a terrier could improve the outcome of our hunting. By combining our personal observation and experience, we arrived at what we regard as the discipline that a rabbiting terrier requires and the training that will encourage this.

It need hardly be said that in order to perform its tasks effectively, the rabbiting terrier must have a combination of well-developed instincts, an active mind and careful training. The simple reason for including training in the programme is because a terrier that is not used to discipline and control is apt to be led astray by its strong will and to 'pursue its own agenda', which, more often than not, is in conflict with the wishes of the handler.

Training is the process whereby terrier and owner develop the team spirit and harmony necessary if their hunting exploits are to bear fruit. The terrier can be a determined little creature with a mind of its own and the attention span of a gnat. They are not immune from thinking that they know best, and can exhibit a woeful degree of stubbornness. However, do not believe for one moment that a terrier of any breed cannot be trained. The oft-heard statement 'But it's a terrier...!' has been used countless times as an excuse for, and even to justify, aggressive behaviour and incessant barking; but, rather than proving that a terrier is incapable of being trained, all it does is conceal the lack of constructive instruction given to the dog. In the course of my reading I came across an account of a Sealyham Terrier that commenced its training when it was the grand age of ten years old, and that same Sealyham later became the star of an obedience exhibition at Rockefeller Plaza in New York City during National Dog Week.

Even so, it is misleading to think that training a terrier is going to be a quick process free from trial and tribulations. At times, terrier owners have been known to shout themselves hoarse and throw their hats to the ground in frustration, and even dog-training experts admit that terriers are amongst the most challenging of breeds to train. The owner should therefore prepare himself by getting to know the character of his terrier, and should develop a training programme that has realistic expectations. It is also imperative that careful thought be given to the method of training that best suits the terrier and handler. I endorse training that is relationship based, which prompts the dog, with the minimum of encouragement, to obey because it enjoys the pleasure and approval of the owner as its reward, sooner than a constant supply of food treats.

Terriers, like most dogs, respond well to copious amounts of praise, especially when they are made to feel that they are being particularly clever. However, inevitably there will be some sessions when your terrier behaves stupidly, ignoring every command and running about wildly – but even in these situations there is still no need to resort to physical punishment if you have taken the time to build a proper friendship with your dog. Working terrier men have always known this, and one writing in the late 1800s stated 'I do not believe in thrashing dogs, or in giving heavy punishments at all. Get your dog fond of you, and when it does wrong, make it feel ashamed of itself. That will be punishment enough.' These are wise words and are well worth paying attention to, and you will find that the voice is much more capable of chastising the dog than the hand.

The only way to approach training terriers is with a great deal of patience and a basic plan to guide you. My brother suggests that their training can be divided into three distinct phases:

- basic obedience;
- advanced training;
- the encouragement of natural instincts.

The combined intention of this training is to:

- provide the terrier with the knowledge and behaviour that is expected of it; and
- provide the handler with the confidence and trust in his dog, which is essential if he wants his terrier to do its work without constant instruction.

Before we look in more detail at these phases of training, I would like to offer some guiding rules which, if adhered to, will make the training sessions a lot easier. There are four of them, and they are as follows:

- **Avoid confusing your dog** with the commands that you verbalize. You will be surprised at how quickly a training session can go badly wrong simply because you are failing to make clear to your dog what is required. An alteration in the words of command, or even the tone of voice in which the command is delivered, can be enough to put the terrier off. Therefore, select your words of command before you start to train your dog, and do not alter them. Do not underestimate the effect your tone of voice has on your dog. The terrier will be able to sense any anger and frustration that is creeping into your words, and this will make it hesitant and possibly distressed as it tries to figure out what it is doing wrong.
- **Make the sessions enjoyable and keep them relatively short**. Fifteen minutes each day is often better than an hour once every blue moon. As with most things, consistency is important, and regular sessions will get the best out of your terrier.
- **Do not lose patience with your terrier**, no matter how sorely you feel you are provoked. An angry trainer who constantly wants to reprimand his dog for every indiscretion will never achieve much. As I have stated, the handler should not resort to using physical force to discipline his dog, and years of keeping a variety of dogs has convinced me that delivering blows to the

rump of a dog in an effort to punish its bad behaviour is a needless and futile way to act.

- **Be prepared to take advice**. If you encounter difficulty with a terrier, on no account give up on the dog. There are numerous people who will be happy to help you, ranging from working dog enthusiasts to dog trainers. A few words of guidance or being shown how to handle your terrier can make a world of difference.

We are now in a position to look at the three phases of training in more detail, highlighting their purpose and describing the techniques that are used to accomplish them.

BASIC OBEDIENCE

The intention of teaching the rabbiting terrier basic obedience is to enable the handler to control his dog. It is the foundation stone upon which everything else is built, and it consists of four commands: 'come', 'sit', 'heel' and 'stay'.

There are some who believe that a working dog should not be subjected to basic training until it has learnt its sport in the open field. This is nonsense – and so, too, is the notion that a thorough training discredits a terrier and robs it of its courage. A terrier that lacks elementary training will be an absolute nuisance if it is taken hunting, and in such situations both handler and dog will rapidly tire of one another. I will, however, concede that basic training is, and can only be, perfected when the terrier is hunting, because only then are the commands put to their most severe test, when the dog is so highly stimulated by the sight and scent of rabbits.

It is also true that you will only fully realize the value of basic training when you begin to hunt rabbits with your terrier. As we shall see, the commands have a specific function when hunting rabbits, and are no longer merely exercises of obedience. My feeling is that basic obedience is more difficult to teach the terrier than all the other aspects of training: this is because advanced training and the stimulation of instinctive behaviour are more closely associated with hunting prowess and therefore capitalize on the terrier's inherent skills.

I shall now take the four commands of basic obedience in turn, and describe the techniques I rely on when teaching a terrier.

'Come'

It is self-evident why this command is of such importance: if you cannot retrieve your terrier, you will not be able to let it off the lead, which makes it absolutely useless as a working terrier of any description. The 'come' is also vital for reasons of safety because it enables the handler to call his terrier away from dangerous obstacles and situations such as electric fences and busy roads.

This command is one of the easiest exercises to teach the terrier, and begins during puppyhood when the gregarious dog's natural inclination is to return to its owner. All that is required is a little encouragement in the form of a coaxing voice, and I start this in the confines of the home where there are no

The terrier must be taught to return to the handler when it is a puppy.

distractions for the dog and only a small distance for its short legs to travel. Once the terrier has got the idea, it is time to venture outdoors and you should be prepared for an element of failure to start with due to the fact that there will be noises, smells and even other animals competing for the attention of the curious puppy. For this reason, I attach my terrier to a long lead and allow him to wander a short distance from me, before calling his name and asking him to come, using the same coaxing voice as I did when inside the house. If I fail to gain the attention of the dog, I can give a gentle tug on the length of leash to reinforce my verbal cues and gently guide him back in my direction. As the terrier improves I abandon my coaxing voice and speak to him in a normal tone, and when I am satisfied that he has reached a suitable standard, I dispense with the long lead. I then go through exactly the same process as when he was on the lead.

As with each new aspect of training, problems may be encountered to start with, and it always pays to have given some thought as to what you can do if your dog forgets what it should be doing. When you first let your terrier off the long lead you may have trouble getting his attention, and for this reason I carry a whistle in my pocket. If my dog ignores me, in spite of his name being called three times, I make a short shrill blast on the whistle, which makes him focus immediately on me and I am then in a position to calmly issue the command. I believe this is preferable to shouting at the dog when it fails to respond and, if you have an inquisitive terrier, the sound of the whistle will encourage him to return because he wants to find out what the strange new noise is. If this proves to be the case with your dog, you should use the whistle in conjunction with the verbal command 'come'. I have found the whistle to be very effective in helping with the recall of my terrier, especially in those situations where there are a multitude of other things he would rather be doing than listening to my voice.

If you have the misfortune to have a stubborn character who will simply not return to you when he is released from the long lead, do not in any circumstances chase after the miscreant. It is much more effective to copy some of the techniques that the horse whisperers of America commonly employ when they are establishing a relationship with a horse. For instance, instead of walking towards a dog that will not come to command, try walking slowly away from him. I have done this numerous times with my terrier, and discovered that when he thinks he is being ignored, or that I am disappointed with him, he cannot resist coming to me. The final option is simply to sit down and just wait for as long as it takes for your terrier to come back to you of his own accord. This is what the native American Indians do with their horses, and it requires the patience of a saint, as it is an extremely slow process.

It is notoriously difficult to call back a terrier when it is in hot pursuit of its quarry, and the inexperienced handler may be shocked as he watches the heels of what he thought was a trained terrier disappear into the distance as it chases a rabbit. However, if you allow your terrier to chase rabbits in a safe environment and give the command to 'come', either when the chase is over, or when the rabbit has opened up a large gap between itself and the dog, you will be surprised at how responsive the terrier will become. Eventually, after practice in these conditions, you will discover that you do actually have the control to call your dog back, even when it is about to launch itself into a chase of the rabbit.

'Sit'

This command is important because it enables the handler to place the terrier in 'neutral', in similar fashion to the command 'stand at ease' used with soldiers. During the course of a day's rabbiting there will be a number of times when you want your terrier to take a pause from its work, for instance when setting purse nets, and the 'sit'

The terrier in the 'sit' position, appearing comfortable and alert.

command is the ideal way to accomplish this. The terrier is not a breed that welcomes inactivity and to start with it will not understand why you want it to take a break by sitting motionless, and you will probably encounter a certain level of resistance. However, take the time to correct the terrier's position and calmly reassert your wishes, and you will soon establish the 'sit' as part of the working habit of the dog.

Before you can expect your terrier to sit on command when you are hunting, you will have to teach him the 'sit' within his home environment. This is one of the easiest commands to teach the terrier, and there are two basic ways of doing it. The first, which is probably the most common, is to apply gentle pressure to the dog's hindquarters immediately after you have spoken the command; after a number of repetitions the dog will be performing the 'sit' without the need of any persuasion. The second way to teach a dog to sit is to take it for a walk on a lead and during the walk, stop and stand still for a couple of minutes. If you remain standing still and wait long enough, the dog will sit down quite naturally of its own accord, and all you need to do is praise him for doing this.

Obedience experts want their dogs to sit in a particular way when they are given the command, but for the rabbiter, such fine points are irrelevant. To be honest, when I am using the 'sit' command, I don't really care on which side of me the dog chooses to sit, or which way he is facing. In fact I am not even bothered if he lies down and rolls on his back, because I am using the command to engage a change in the dog's attitude which, in terms of working, is of more importance than the outline it assumes.

Once the terrier is accomplished at sitting near you when it is on and off the lead, you should teach it to sit on the spot when it is some distance away from you. You should give a crisp verbal command and rely on your previous training to achieve this. Again, with time and repetition, the terrier will soon become an expert. You can, of course, vary the use of the 'sit' command to suit the way you like to work your dog when you are hunting rabbits. For instance, my brother only used the command with his old Jack Russell when he wanted to signal to her that her work was finished for the day so that she would calm down after her strenuous activity in the same way that an athlete cools down after running a hard race.

'Heel'

The command 'heel' enables you to take your terrier through farmyards and other potentially hazardous areas without it getting in the way of tractors, or being a nuisance to flighty livestock. A dog that follows you like a drunken sailor is likely to embarrass you as it meanders off in all directions. For those who wonder why the terrier is not simply put on the lead when going through farmyards and past livestock, the answer is straightforward. It is impractical when weighed down with a dozen rabbits, ferret box, spade and game bag full of nets, to put everything down in order to place the terrier on the leash; and if you have to keep your dog on the lead to get through livestock, you are going to have an impossible time rabbiting on cattle and sheep farms.

Unless released to carry out its work, the terrier should always be at heel, and even the sight of a rabbit should not pull it out of 'heel' until permission is given. The importance of this cannot be overemphasized, because terriers that are undisciplined and wander hither and thither as they see fit will ruin the day's sport and, even worse, can be the cause of an accident on a busy farm. A dog that behaves like this will quickly earn you a bad reputation, and farmers are understandably reticent about allowing unruly dogs on their land.

The 'heel' command is also of relevance to those situations when it is absolutely necessary to put the dog on a lead, such as crossing busy roads, and it is also used to prevent the animal from continually pulling on the lead. This makes walking with a load on your back much easier, and stops the dog getting on your nerves with what must be one of the most annoying forms of canine behaviour.

If you observe the competitors at obedience trials, you will notice that their dogs are trained to walk at heel on the left side of the trainer's body, and they are taught to do this for several reasons. The most obvious is that the majority of us are right-handed and suffer less interference from, and have more control over, a dog when he is made to walk on the left side. Originally the purpose of such training was to enable the right hand to be free when the dog was used for police work. The prisoner would be made to walk on the policeman's left side with the dog in between them, ideally placed to pounce on the prisoner if he reached for a weapon of any kind.

The dog must first be taught to 'heel' when he is on the lead, because in the initial stages the lead will be used as an extra method of communicating with him. First, give the spoken command clearly to your dog, take a brisk step forwards, and give a gentle tug on the lead to reinforce your wishes. Your dog should then walk beside you without pulling, and be given praise accordingly. If he walks ahead of you, the command 'heel' should again be used, but this time with more firmness and a quick tug of the lead to check his pace. As he comes into line with you, he should again be praised. It is then simply a matter of repeating the process until it is acquired behaviour.

If your dog is unmanageable, one expert recommends that you turn and walk in the opposite direction every time it lunges ahead. A series of fast about-turns, one after the other, will apparently get results more quickly with an unmanageable dog than anything else.

Once the dog is fully versed in walking to heel when on the lead, it is time to teach him to 'heel' when he is free. When this is accomplished, the handler will be able to trust his dog to walk close by his side and will be secure in the knowledge that the dog is under complete control. Begin by holding the leash so slack that it places no restraint on the dog: this can be done by draping it over your shoulder or tucking it into your belt. If the dog gets out of step, use your voice to correct him and only use the lead in emergencies, or if you feel he needs reminding of his former lesson. When you are satisfied that his movements are being controlled entirely by your voice, you can progress to taking the leash off and have him walk to heel in response to your spoken commands. Regular sessions will soon perfect this exercise, and the command will prove its usefulness in everyday life, as also when hunting.

'Stay'

This command is used to place the rabbiting terrier at a certain location in a state of alertness, such as by a bolt hole that cannot be netted. It is not intended to place the terrier in 'neutral', as with the 'sit' command, and the terrier must be made to realize that when the 'stay' command is used he can move when he deems the time is right in order to pursue or capture quarry. Once placed in the 'stay' position, the terrier should be trusted and not interfered with, even if the handler thinks

The terrier in the 'stay' position, but ready to respond if a rabbit bolts.

that he is making his move too early. It is well worth remembering that dogs' senses are far more acute than ours, and a dog can anticipate when a rabbit is about to bolt with far more accuracy than we can.

The way to teach the 'stay' is to take your terrier to a rabbit warren that you know to be in regular use, and *do not forget* to take your ferret along, and also an assistant. Place your terrier above the warren and tell it to stay, which of course it will not at this early stage of training; this is why you should take an assistant with you who can restrain the dog and, at the same time, make it comfortable. When the terrier has been settled for a couple of minutes, the ferret should be entered into the small warren and the terrier held in the same position.

Having omitted to place purse nets over the bolt holes, the rabbit should make a hasty exit from the warren, and as it does so the terrier should be immediately slipped from his lead so that he can chase the rabbit. Your terrier may be quick enough to catch the rabbit as it bolts, but this does not really matter at this stage. Repeat the process a number of times, and then when you feel the terrier is ready, transfer to voice commands only to hold the terrier in position.

When this has been accomplished, you will be able to effectively guard holes that you don't want to, or cannot, net. I think that this is the best way to teach the rabbiting terrier that the movement of the rabbit is as acceptable a trigger for the end of the 'stay' command as the handler's voice. In all honesty this is a confusing lesson for dogs, but terriers have an instinctive feel for working warrens, and soon learn what is required of them.

Someone may well ask the following question, 'If I use the "stay" command in this way, how can I make the terrier remain in the same position for a prolonged period of time without it moving?' My brother achieved this with his Jack Russell bitch by using the command 'down and stay'. This may be only a subtle difference, but when it is combined with the different environment in which the command is used, the terrier soon realized what was expected of her. She became legendary at performing this task, and we could walk the length of a substantial field with her remaining still at one end of the field while we approached the other end.

My brother taught her this simply by using his voice and some hand signals. He started close to the terrier, gave the command 'down and stay', gave a supporting signal showing the palm of his hand, and then moved a metre away from the dog. He gradually increased the distance until his terrier became a mere spot on the horizon. When conditions would not allow his voice to travel, he was able to communicate with his dog by using a hand signal. It may also be worth mentioning that when he used the term 'down', he was not really bothered which position she assumed, as long as she was not moving.

ADVANCED OBEDIENCE

Having looked at the basic training of the rabbiting terrier, we can now turn our attention towards advanced training, which basically means teaching the more specialized skills that a terrier will require if it is to be a successful hunter. To teach the terrier these skills, you have to probe the psychology of the canine mind in order to know how to inspire your dog to move up a gear. My brother suggests this by stating that, in order to train a terrier, you have to think like a terrier. The main objectives of the advanced training programme are:

- to conduct the rabbit hunt in a disciplined fashion;
- to make the hunting of rabbits safe for the terrier;
- to ensure that the rabbits are hunted in a humane way.

We all know from what has happened to fox hunting that the humane angle of hunting has become a hot potato, and the last thing that you want is a terrier that operates in what may be perceived to be a barbaric fashion. Terriers need to be efficient, disciplined, and above all, controlled killers of prey – but do not go off with the idea that this objective is driven by the need to look good in the eyes of onlookers. It has always been my opinion, which is shared by the vast majority of field-sport enthusiasts, that a hunter – regardless of whether he is pursuing a fox, a rabbit or even a rat, has a moral obligation to make the death of the quarry as quick and painless as possible. The native American Indians maintained their respect for the animals they hunted because they viewed them as gifts from a spiritual benefactor. The rabbiting terrier adheres to no such notions, and it is a hard task to control the blood lust of an adrenalin-pumped terrier: but with experience, it can be done, and it is made much easier if the terrier has a thorough grounding in advanced training.

The advanced training programme that follows is divided into four elements:

1. Stock familiarization
2. Food removal
3. Handling and scruffing
4. The development of natural instincts

When you have taught your terrier to perform these four tasks proficiently, you will have a terrier that is ready to hunt.

If you start to work your terrier before it has mastered the principles of advanced training you will be putting it at risk and exposing the quarry to possible inhumane treatment, and will thereby be laying yourself and the sport open to criticism. Furthermore, a terrier that has not received advanced training is unpredictable, and on a farm where you could encounter anything from young lambs to flighty cattle and even chickens, such a terrier could prove to be a dangerous liability. If it should run wild, you will certainly be forbidden to hunt over that land in the future, and may even face the threat of prosecution if valuable stock is damaged. This suitably highlights how vital advanced training is for the sporting terrier.

Before we begin to look at the elements of this level of training, I would like to draw your attention to the amount of time that it will take to instil this degree of obedience into the wooden head of your terrier. It will take many hours, spread over several weeks, as you build brick by brick, and you should not expect immediate results, however much effort you put into it. Some days will go so well that you will herald your dog as a genius, and other days will be so appallingly bad that you will think him a cloth-eared fool. Do not despair when this happens because it is a natural process that you have to go through with your dog, and everybody who has trained a terrier to an advanced level has encountered exactly the same problems as inevitably you will. A competent terrier man is the product of much study, and hours upon hours spent in the company of terriers.

Stock Familiarization

By definition, this is the ability of the terrier to differentiate between animals that it is allowed to hunt and those which it should never threaten under any circumstances. It also involves teaching the dog the manners that it should display when it is in the presence of livestock.

My brother's old terrier would quickly kill any rabbit that she could get her teeth into, and yet when we lived in Cambridgeshire, we owned the largest and most mischievous lop-eared rabbit you have ever seen, known as Gilbert. Gilbert was permitted to live in the house, and not only did his terrier allow him to move around the house freely, she even tolerated the outrage of the cheeky rabbit sleeping in her dog basket. The terrier's natural inclination was to kill Gilbert, who was, after all, a rabbit, and she was a typical working terrier with all the vermin-killing desire you would expect. However, something stronger than her natural behaviour was at work, which kept her in check, and this was her deference to the training she had received. This gave her the capacity to discern pets from legitimate quarry.

My Jack Russell dog also possesses this gift. We were made aware of this one day when we were after a rat that had somehow managed to get itself into a barn wall behind a large water tank. We sent in a workmanlike little polecat hob who was soon on the trail of the rat, and the next thing we knew my Jack Russell, called Slipper, had dropped to his belly and crawled in after the ferret. For a moment we were a little concerned and wondered whether Slipper would, in the heat of the moment, recognize the difference between friend and foe. It did not take long for us to find out, as the terrier and ferret were soon nose to nose, but they both ignored one another and continued to pursue the rat.

It is vital that you teach your terrier there are certain animals it may pursue and others that are strictly off limits, otherwise it will

The terrier must get along with livestock. Here my terrier is playing with a young billy goat.

kill anything it likes including pet rabbits, cats and even your ferrets, given half a chance. Some time ago I had a friend who had a lovely collection of small workmanlike ferrets; he also had a terrier that had not been given any advanced training. One day he was working with his ferrets when he was called to the house for a moment. When he returned, he discovered that the terrier had jumped into the ferret run and killed every single ferret. Such behaviour cannot be entirely blamed on the terrier, but must be shared by the owner who failed to provide the dog with a sufficient level of canine education.

I have learned that exposure is essential, by which I mean that you should expose your terrier to as many different animals as possible. Older terriers can also be trained in this way, but it will probably require a handler who can be firm and has plenty of experience of training terriers. You should start by taking your puppy into the ferret shed and allowing it to sniff at the cages; however, *don't* let it stick its tongue through the wire on the front of the cage, as my reckless terrier did when he was younger. The result was predictable, as the ferret seized hold of the end of the tongue and pulled vindictively as hard as possible. It was fortunate that the terrier was not put off ferrets for life.

After a few days of allowing your terrier to sniff around the ferret cages, you can move on to the next stage, which is to introduce your terrier to a big, playful and, above all, friendly ferret. If your puppy is friendly towards the ferret offer it plenty of praise, but if it displays any signs of aggression, instantly scold it with the voice. Once your terrier has been well socialized with a ferret, you can then introduce it to any other animals that you may happen to keep, such as chickens, cats or rabbits. Remember to check bad behaviour immediately, but equally to reward good behaviour just as rapidly.

At the same time as you are teaching your terrier not to harm domestic animals, it is a good idea to walk him over some ground that is rich in wild rabbits, which he should be allowed to chase freely, receiving plenty of praise and encouragement for his efforts. The reason for doing this is that the terrier needs to be taught that it is acceptable to pursue animals as long as *you* have given permission. If you were to neglect this part of its education, the terrier would become unbalanced and in its mind would think that all animals come under your protection, whether they be rats, rabbits or chickens.

When you have accustomed the terrier to your own pets, it is time to visit a farm and introduce him to pigs, cows, sheep and horses. This should be done with the terrier on the lead at first, and any aggression or excitability should be checked with the voice and a tug on the lead. A terrier that cannot be broken to livestock, no matter how promising a worker it may be, will always be a liability as no farmer with any sense will allow it on his land; consequently, perseverance in this department is essential. When your terrier will go through stock quietly whilst on the lead, you will be ready to take it through at 'heel' and off the lead. As long as you have spent a sufficient amount of time on the former exercise, you will have a fairly accurate idea of how your dog is going to react when he is free, and so you should be able to avoid any serious problems.

I have always found it quite easy to elicit the correct behaviour from our dogs when they are in the company of farm animals, and it should be remembered that a number of terriers, particularly the Irish ones, were bred to be all-purpose farmers' dogs. Terriers have more trouble accepting small verminous creatures, but with time and association, they will come to realize that even these are not their enemies. Do not, on any account, expose your puppy to farm stock that is overly aggressive. I have known of horses and cattle that will attack dogs if they are given a chance, and if a little terrier is subjected to an attack by such a Goliath it will, understandably, have difficulty trusting any form of farm animal in the future.

Removal of Food

The intention of this aspect of training is to teach the terrier to release a rabbit that it has caught when the handler gives the command. The reasons for this are to prevent damage to the carcass and to ensure that the rabbit is killed as quickly as possible. If you fail to teach the terrier this discipline, it will continue to maul the rabbit after it is dead and will lose sight of his task. In those rare situations where the terrier only manages to grab hold of a limb of a live rabbit, you will want the dog to release its grasp instantly on command so that you can dispatch the rabbit without delay.

My brother's old terrier was not constrained by any moral code when she was hunting, and would grab a rabbit by whatever was available. We have seen her take rabbits by the head, around the rib cage and by a hind leg: for those grasped by the head or the rib cage, death is almost instantaneous; however, if the rabbit has been taken by the hindquarters, death will be much slower in coming, and will involve pain and fear. On such occasions the terrier man has to act swiftly, removing the rabbit from the mouth of the dog so as to effect a rapid kill. It is not natural for a terrier to give up its quarry, and this quality must therefore be learned – and the place to begin is with the removal of food.

Dogs as a rule do not like to give up their food, and some will even bite if you go near them when they are feeding, but my brother's terrier will allow him to remove her bowl whilst she is feeding and even allow him to take the food out of her mouth. Such unnatural interference can be made acceptable to your terrier by removing things from it when it is a tiny puppy, and continuing with the activity until the principle is well established.

When my brother's terrier was a puppy, many moons ago, he removed her food when she was eating and took biscuits and toys away from her on numerous occasions. He always praised her liberally for allowing him to do this, and then returned what he had taken. The result of this training was that he could walk up to his terrier when she had a rabbit in her mouth and remove it without her procrastinating.

If you have an adult terrier that has not been trained in this technique and it displays an obstinate reluctance to release its grip when it is hunting, I would suggest that you modify this method slightly. This is because I do not believe it is fair to disturb an adult dog that is totally unfamiliar with interference when it is feeding. It will be more appropriate to employ an old rabbit skin. Begin by tempting the dog to grab hold of the skin, and if your terrier is anything like mine it will not require much tempting. When the terrier has hold of the skin for a few moments, you should then ask him to release his grip with the command to 'leave it'. Repetition of this process will usually lay the basis for success when you issue the same command within the context of the hunting situation.

Handling and Scruffing

This essentially means the ability to handle your dog in a completely unorthodox manner without it offering any resistance or harbouring any resentment at the way it is treated. Terriers used for rabbiting are by necessity little creatures that in the course of their work go to ground, or, as with my terrier, when I am located at a derelict farmhouse, disappear under the floorboards. This combination means that they sometimes get themselves into tight places from which they find it difficult or impossible to extricate themselves. When this does happen you need to be able to grab whatever part of your terrier you can reach and pull. If your terrier has not been trained for this, it will tend to tense up and become stressed to such a degree that it may try to bite you.

When a terrier becomes tense its body size increases very slightly and it will be less pliable, and a dog that gets itself stuck may panic and can be very difficult to assist.

In order to get a terrier used to rough handling you need to start when it is still a puppy. Begin by scruffing it, which simply means picking it up by the loose skin around its neck in a similar fashion to the way its mother would lift it. Next, when your terrier is rummaging around the house, wait until it goes behind a chair or better still under a bed, and then extract it by its hind leg, its tail, or the loose skin on its rump. When you are doing this, take a gentle grip and apply constant pressure as you slowly pull the dog towards you. The disposition of most terriers will mean that this process will soon become a game to them. It may sound a bit cruel, but believe me, this training can literally save the life of your terrier. To stress this point, I shall give you a few examples from experiences that my brother and I have had.

Being able to handle the terrier roughly is sometimes essential.

My brother was out on a walk one day when his terrier suddenly picked up the scent of a rabbit and disappeared below ground into a small rabbit burrow. He waited patiently for a rabbit to bolt or for his terrier to return, but after ten minutes of inactivity he began to get a little concerned. When he looked into the warren he saw a perpendicular drop, with his terrier just visible by her nose at the bottom of it. He called her and she did try to come forwards, but slipped back each time like a car on ice. Heavy rain the night before had turned the drop into a slippery skid pan. My brother lay on his belly and, reaching as far down the hole as he could manage, was just able to touch her wiry little whiskers. Helping as best as she could, she moved forwards all of an inch and pushed her snout into the palm of his hand. He gently closed his fingers to give him a firm grip and pulled his terrier to the surface by the end of her snout. On another occasion, he managed to pull a friend's drowning terrier from the water by its scruff just as it disappeared from view. I have myself used the same hold of the scruff to retrieve my terrier from the small gaps between round bales of stacked straw. Undeniably, scruffing and rough handling may not look pretty, but it is of vital importance for the working terrier.

The Encouragement of Natural Instincts

It could be said that this basically means teaching the terrier to hunt rabbits, although I would argue that it is just as much about placing the terrier in an appropriate environment that provides it with the initiative to hunt. This element of training makes the terrier aware of what nature has given it, and gives it the confidence and knowledge of how best to use its inherent talents.

By far the easiest and most successful way to make a terrier realize what it can achieve when hunting is to partner it with an older, 'tried and tested' terrier. This is, after all, the way that animals are educated in the wild. If you do not have a tutor in the form of a more experienced terrier, then it is up to you to help develop its natural predatory behaviour.

You should begin by taking your terrier to a field where you know it has the opportunity to bolt rabbits from seats and other temporary shelters. Your prior knowledge of the ground should enable you to take the dog to the most frequently used hiding place of the rabbits. When you have reached this position, take a pause and allow the terrier time to sniff the ground carefully. As it picks up a scent you should offer verbal encouragement and watch the dog follow the scent until hopefully it bolts a rabbit. If your terrier has difficulty following a scent, I would suggest that you take it slowly over the land in a variety of directions until the scent is strong enough to grab the dog's attention.

After a few outings your terrier should be able to discern the pathway of a scent trail with some accuracy, and should be guided to follow it to a warren. You can then use that warren like a classroom, where the terrier can hone its marking ability. All that is really required of the handler is to give the terrier copious amounts of time, watch carefully for the desired behaviour, and praise the dog suitably when it exhibits it.

I have heard people complain that their terriers have no idea what to do when they are first taken hunting – and this is not surprising if time has not been spent educating and developing the mind and instincts of the terrier. I have discovered the harsh truth, that if you want a better terrier, you must first become a better trainer, with a more intensive programme of instruction. I would never make the claim that my approach to training will make the terrier a first-class hunter, but it will provide it with a knowledge of what the handler requires. The owner will also be confident that he can control his dog, and will therefore be relaxed enough to let the terrier find its own feet when it is taken hunting. Time and again you will make use of the commands that you have taught your terrier during the phases of basic and advanced obedience training, and you will be pleased that you bothered to invest the time and energy in him.

The young terrier being taken on unusual ground in order to develop his confidence and sense of smell.

CHAPTER SIX

The Health and Welfare of the Rabbiting Terrier

This chapter has been contributed by my brother, who has studied welfare issues with regard to working terriers in far more depth than I have. He writes regularly for *The Countryman's Weekly*, in which he published a long-running series dealing with the subject of the elderly working dog.

THE DEMANDS OF A RABBIT HUNT

To start this topic we need to think about the nature of a rabbiting terrier's duties in relation to the physical and mental demands they place upon him. The moment he arrives on the ground to be worked he starts to use his nose and, in order to analyse the scents he picks up, he will have to engage his brain. A day's rabbit hunting will last between four and six hours, and for a good part of this time the terrier's nose is active. At the close of the day you will therefore end up with a terrier that is mentally exhausted.

Mental exertion fatigues the dog's body just as surely as physical exercise, and the circumspect owner should realize that the mental demands placed upon the rabbiting terrier eventually exhaust him; this means there will be a limit to what he can and should endure. Furthermore, the younger the dog is, the more quickly its mental powers will be tired out. This is one reason why a terrier is not up to a full day's work until it is eighteen months old, and no terrier should work for more than six hours continuously. It is best to divide the six hours of work into three-hour sessions, with a midway break of twenty minutes' duration. In those locations where the work is particularly demanding I would recommend that the terrier's exertions be confined to a period not exceeding three hours.

You might not think that your terrier is slowing down, but mental weariness will show itself in the dog when it gets back home. The terrier will sit in its bed or in front of the fire with eyes staring vacantly and obviously wanting to go to sleep, but unable to do so because its mind is still switched on. In this peculiar state the terrier will wander from one place to another, settling for only a moment before moving on again. Eventually the worn-out tyke will make itself comfortable and drift into a deep sleep that should not be disturbed even if it lasts for a couple of hours. The only remedy for lassitude is rest, and plenty of it. As you read through this chapter you will soon realize that rest is as important to the health and well-being of your dog as good nutrition, correct housing and adequate amounts of exercise.

After a hard day's work, the terrier requires a good rest.

In the course of a day's hunting the terrier will also have to move at speed, sometimes in open spaces and sometimes in thick covert, where he has to jump, dodge and weave to get through the undergrowth. When he goes below ground or into some huge woodpile or suchlike, he will twist and contort his body to its extreme limits so that he can pursue his quarry through incredibly tight tunnels. High speed pursuit and subterranean work place enormous strain on the terrier's major physical system, and during the course of a day's work the dog's muscles will be stretched to the absolute limit; they must feel as if they are being torn to shreds.

Working so close to the limit of its physical abilities lays the terrier open to the possibility of injury to its muscular system, and at the end of a day's rabbit hunting the terrier will probably display some degree of stiffness as a result of its strenuous exertions. This is nothing to be concerned about, and again rest is the recommended treatment in conjunction with warmth. If the stiffness should persist into the following day, the terrier should be offered some light exercise to help loosen the muscles, but you should avoid exposing him to wet weather. You can also massage the affected area by moving your hand in concentric circles on the stiff limb, and a heat pad may be placed on the relevant area while the terrier is resting (the pad should be checked with the back of the hand or elbow to ensure that it is at the right temperature).

It can clearly be seen that if the rabbiting terrier is to be fit for his arduous duties he must possess an abundance of strength in his bones, muscles, tendons and ligaments, and stamina facilitated by a powerful heart and pair of lungs – and all this must be combined with adequate maturity to cope with the physical and mental stresses involved in hunting.

THE AGE TO START WORK

Because dogs are naturally fit and healthy animals, we tend to make demands on them without really thinking about any long-term consequences. The man who works terriers must realize that there is a direct correlation between the strenuous activity his dog undertakes and his health. Due to the influence that hunting exerts on the health of the terrier, we must seriously consider what would be a suitable age for him to commence his sporting endeavours.

The young terrier must be given time to grow and mature before being considered for work.

Starting Work Too Early

I knew of a gamekeeper who had a black Labrador which he started to work in earnest too early in life, and by the age of four the poor creature was showing all the classic signs of advanced age. The internal organs of the dog were almost worn out, it suffered from arthritis, had a depletion in muscle mass and its coat was dull. This dog was also affected mentally and had lost the will to live. By the age of seven, when it should have been in the prime of life, it was dead.

This unfortunate Labrador had been worked into an early grave by the ignorance of its master, who placed demands on the dog that were too heavy too early in life and which precipitated a process of steady degeneration (this means a change in the structure or in the chemical composition of a tissue or organ by which its vitality is lowered or its function impeded). This process can result in brittle bones, impaired kidneys, a damaged heart and a plethora of other avoidable conditions. The startling conclusion is that working a terrier hard too early in life can quite literally destroy its health.

If you were to go outside and try to push down a mature oak tree, it will obviously not yield to your efforts; but if, on the other hand, you find yourself a young sapling, you will be able to snap it in two with ease. The young terrier is like the oak sapling, with all the structures and organs of his body being tender, and if he is subjected to inappropriate levels of physiological stress, a degenerative process will be triggered.

Bone Development and Starting Work

It would be easy if the terrier could wake up one day and tell you that it is fit and ready for work. In a way this is exactly what happens, but instead of using words, the terrier can indicate its readiness for sport by the stage of development of its bones. This is not as ludicrous as it may sound, and the importance of bone formation is a topic that has been of constant interest to Masters of Foxhounds and owners of greyhounds used for racing and coursing; and it is also of use to our study.

There are approximately two hundred bones in the body of a dog, and their essential function is to support the body and enable it to move. The bones in a young dog's body are not as lifeless as many people think, and are in fact very vibrant living structures that are growing and changing all the time. During

A Guide to Age-Related Activities for the Rabbiting Terrier

Activity	Age to Start	Notes
Basic obedience	4 months old	Very short training session of about five minutes
Advanced obedience	4 months old	The sooner the terrier starts to mix with stock the better
Stock familiarization		
Advanced training	6 months old	The terrier has sufficient mental and physical maturity to begin light rabbiting. A half day over easy ground in the company of a more experienced terrier is recommended
Light rabbiting	1 year old	
Full rabbiting	18 months to 2 years	At this age the body is fully developed and the terrier is now ready to begin his work in earnest; this will enable him to perfect his training

exercise the bones are subjected to a surprising number of physical pressures, such as shearing, bending, twisting, lengthening and shortening. The size and strength of bones will increase in response to exercise, but if the dog stresses itself with movement that is too taxing, the bones will be weakened because the cells that promote growth cannot rebuild as quickly as the damage is caused.

The skeleton provides the foundation for the working terrier's physical structure, and it is therefore essential that we wait for the bones to be at the appropriate stage of development before we start working the terrier. The bones of the skull and spine can take up to two years to reach maturity; those of the leg develop slightly more quickly and are usually fully developed between twelve and sixteen months. Consequently, when the terrier is twelve to fourteen months old, it may be given a taste of light work, but should be held back until its second year before it is allowed to exert itself to the utmost of its physical powers.

At one year old the terrier may take part in rabbit hunting, however this should not be for a full day or in surroundings that are too demanding, and it is best to work the young dog in the company of another that is more experienced. You should not ignore or treat with contempt the guidance that bone development offers, because over-exercise will permanently weaken and stunt the growth process, whereas an appropriate amount of exercise that increases in intensity in response to age, and a correct diet, will give the bones the resilience to support the dog throughout its entire working life.

BUILDING STAMINA IN THE WORKING TERRIER

The terrier man should know how to keep his dog fit, and in order to perform its hunting duties the terrier must undergo endurance training that will develop the necessary stamina. Stamina can be defined as the ability to sustain physical output for extended periods of time, and is basically dependent upon the ability of the body to burn fuel.

The fuel used by the dog is in the form of sugar that is stored in the muscles as glycogen and is present within the blood as glucose. In order to utilize this raw energy, it must be combined with oxygen, and the greater are the physical demands, the more oxygen will be required. This oxygen is breathed into the lungs, absorbed by the blood passing around the lungs and pumped rapidly to the muscles by the heart.

Consequently, stamina in the dog will be relative to the condition of the heart and lungs, which can both be radically improved by training. Training will increase the lung capacity of the terrier and improve the transport of oxygen to the muscles, which are then able to burn more fuel and provide the dog with more energy and stamina. Training can increase the size of the heart by as much as one third, and improve its ability to pump oxygenated blood to the target areas during prolonged periods of exertion.

The rabbiting terrier is without doubt required to be a canine athlete, and to help him achieve the necessary fitness you must follow some form of physical training schedule. With this in mind I have devised the following programme as an example.

TRAINING PROGRAMME FOR THE RABBITING TERRIER

This relies on the performance of four activities: walking, object chasing, grip work and jumping.

Walking

Walking the terrier on a lead should not be the dawdle that many dog walkers seem to adopt. We need the terrier's heart and lungs to increase their work rate, which means that lead-walking should be brisk. The handler should stride out as fast as possible so that the terrier has to trot to keep up. The chart shown opposite (*see* p. 121) highlights the recommended duration of these brisk walks. This exercise will stimulate the growth of strong bones and muscles, and improves the aerobic capacity of the cardiovascular and respiratory systems.

Object Chasing

Every day you should try to get your terrier to have a flat-out sprinting session. You could do this by running with your terrier, but since we require a full five minutes' sprint work it is doubtful whether you would be able to keep up for long. It is much better to have an object, such as a ball, which when you throw, the terrier chases at full speed. Object chasing will quicken the terrier's speed and help with co-ordination.

Grip Work

This basically involves the use of an object, such as a length of rope that your terrier is keen to latch on to with his teeth. Once you have persuaded your terrier to grip the object, you should tug on it firmly. The idea is not to pull it away from the terrier, but to get the dog to resist with all its might. Most terriers love this kind of activity and will give it all they have got without requiring the least bit of encouragement. This resistance training will develop strong muscles in the jaw, neck and back, and it uses up an enormous amount of energy.

Jumping

All you need for this is a series of little jumps of no more than 45cm (18in) high, set up in your back garden. It may take you a little while to teach your terrier to jump, but persevere, because he will eventually pick it up, and jumping is an excellent form of exercise. There is nothing else that is quite so effective when it comes to developing dynamic power in the hindquarters. You do not need the terrier to sprint from one hurdle to another if he does not want to, because it is the jump itself that provides the workout on the hind limbs.

FEEDING THE RABBITING TERRIER

The diet described below may be unlike any other for dogs that you have encountered or used. It is designed to mimic the type of diet consumed by wild carnivores, such as the

FITNESS PROGRAMME

Terrier Aged 4 Months

Day	Morning	Midday	Evening
1	10min lead walk each day	5min ball-chasing 1min grip work	15min walk on lead or 30min free walk each day
2			
3			
4			
5			
6	No exercise	1hr country walk	5min grip work
7	No exercise	1hr country walk	5min grip work

Terrier Aged 6 Months and Above

Day	Morning	Midday	Evening
1	20min lead walk each day	10min ball-chasing 5min grip work	45min–1hr country walk
2			15–20min jump work
3			As day 1
4			As day 2
5			As day 1
6	No exercise	At least 1hr country walk, preferably 2	No exercise
7	No exercise		No exercise

wolf, working on the assumption that nature knows best. The wolf eats the entire carcass of its prey: flesh, fur, stomach contents and bone are all consumed. If we are to copy nature's diet, we should provide our dogs with all the constituents that would be found in the animal that is preyed upon. The wolf preys mainly upon herbivorous animals that feed upon vegetation and whose bodies are composed of the following components:

- Water – most animal bodies contain approximately 60 per cent water.
- Protein – found throughout the body but mainly in the muscles and flesh.
- Vegetable matter – found partly digested in the stomach and intestines.
- Minerals – found in the liver and bones.
- Vitamins.
- Fats – found within the flesh and around the vital organs.
- Fibre – the fur or hair is a source of fibre.
- Sugar – found within the muscles and liver.

We shall now take a more detailed look at these components in order to ascertain why they are important contributory factors to the health and vitality of the dog. A dog literally is the product of what it eats, and this is why a basic understanding of canine nutrition is an essential requirement for the terrier owner who wants to make the most of the working potential of his dog.

Water

A dog can survive for weeks without food, but when completely deprived of water it will perish within days. Water is probably the most important single component of the dog's diet;

it is one of the key elements of life and as such its provision has to be attended to with great care. All the tissues and cells of the body of the dog have water as their basic constituent. It is estimated that approximately two thirds of bodyweight is water, which, in a Jack Russell weighing 5kg (11lb), would convert to about 3kg (7lb). This is clear proof how important water is.

We should remember when providing the dog with water, that he will not view it as an odourless liquid as we humans do. A dog's nose can identify a particle in the ratio of one part per trillion, and therefore it has the ability to smell water. Horses can do this too, and it was recognized by horsemen not so long ago that horses prefer natural water to that which comes from the tap. I have carried out my own experiments and found that my dogs also have a preference for natural water, and would therefore recommend that the terrier is provided with this type of water to drink, and it will not cost you a penny.

You can quite easily collect your own supply by attaching guttering to a shed with a barrel beneath, and with the compliments of the British weather you will soon have a plentiful supply of natural water to give to your dog. However, you should keep a close check on the water during the hot summer months because the increase in temperature can cause vegetable matter to grow, which will stagnate the water that is in the collection barrel. I cannot prove that a dog given natural water is healthier than one that drinks tap water, but there is plenty of anecdotal evidence to suggest this is the case. It is also interesting to note that a dog washed in natural water has a softer coat than one washed in tap water.

Protein

Proteins are complex organic compounds found in both animal and vegetable matter, but the protein found in vegetables is not as valuable as that found in animal matter. The reason for this is because it contains more of the essential amino acids that make up proteins than those found in vegetables and cereals. Only one type of food contains all the essential amino acids that are necessary in order to maintain life, and that is milk. If this were not the case, the young puppy, which is totally reliant on a milk diet, would fail to thrive and grow.

Protein is essential for the growth of young dogs.

The function of protein in the diet is to promote growth and enable the repair of damaged tissue. This is why body builders take it in such large amounts, and why the diet of the puppy must have the correct amount of protein so that it can grow in bulk and develop in strength. Foods that offer proteins are milk, fish, poultry, meat and eggs.

Milk: This is the finest source of protein available, and the best milk of all is whole goat's milk because it is easier to digest than cow's milk due to the smaller size of fat globules found in it. This makes it especially suitable for the puppy, the sick and elderly dog, as well as the working terrier that is in the prime of life. The provision of goat's milk will help to build strong bones and dense muscles, and gives a boost to the immune system, which is responsible for fighting infection and disease. I have used goat's milk taken directly from my mother's small home dairy in the raising of both puppies and ferret kits. The rate of development and the structural strength of these young animals was phenomenal, and because of this I use goat's milk as the major source of protein for all my dogs. If you have the opportunity to read many of the dog books published during the last two centuries, you will find that the authors unreservedly recommend the consumption of goat's milk.

Fish: This is another very good source of protein and has the benefit of containing a substance known as Omega 3 that helps to maintain a healthy heart. Tinned fish, especially pilchards, contains a large amount of calcium and is free from the 'E' additives commonly found within canned dog food. Our goal is to provide the terrier with a varied diet that contains only clean, natural ingredients without the pollution of colorants and additives.

Poultry: Although poultry products will provide the dog with protein, I avoid using them because of the widespread use of growth promoters and reliance upon antibiotics, unless organic products are purchased, and these are far too expensive for regular use.

Meat: I do not give my dogs red meat due to an alarming report I read in a medical book. The report discussed the speed with which meat moved through the digestive system, and recommended the consumption of fibre in conjunction with red meat so that it would prevent the meat being in contact with the intestines for too long. The problem with red meat is the presence of the substances that give rise to cancer, and although the risk to the dog may be small, it is one I would wish to avoid.

Fresh wild rabbit is, in my opinion, the finest meat that can be given to your dog, especially when it is fed raw and on the bone as nature intended. Rabbit fed in this way provides the dog with protein and minerals and plenty of jaw exercise, which helps to keep the teeth strong and clean. As wild meat, rabbit is totally free from chemical interference and is therefore a healthy, lean meat that is low in fat. This is useful, because many of the terrier breeds are prone to becoming rather plump when fed a diet that is too rich in fat.

Eggs: Like milk, eggs provide a form of good quality protein, but they should be boiled so that the dog can easily digest them. If eggs are fed raw, they will cause the dogs to have loose droppings. They could also be fed poached or scrambled with a little milk, and dogs love them. Eggs also contain vitamins and minerals, and could play a much larger part in the diet of the dog than most owners realize. It is interesting to note that farm dogs, like my collies, will eat an entire egg including the shell if they are lucky enough to find one outside in the chicken shed.

Vegetables and Fruit

You may think that you should feed your terrier meat, and meat alone, because he is a carnivore, and it is, after all, the natural diet of such an animal; however, the fox will devour the stomach of a rabbit that is full of part-digested plant matter that the rabbit has picked from the pasture and hedgerow, and foxes have also been observed eating windfall apples, plums, pears and blackberries in the autumn, strawberries and gooseberries in the summer. Records show that they will also eat vegetables such as tomatoes, cabbages, leeks, marrows, peas and carrots. Consequently, the inclusion of fruit and vegetables in the diet of your dog is perfectly natural, and further, is in line with the latest veterinary thinking to come out of America, which recommends that dogs be given fresh vegetables, fruit and grains on a daily basis.

Herbs: Herbal practitioners have observed that many of the diseases that occur in domestic animals do not appear in the wild, and they believe that the reason for this is that domestic animals do not have access to the herbs they require for good health. This argument, it has to be said, does seem to make sense, especially in the light of the fact that whenever dogs are given herbs in their diet, they readily take them as if instinct is telling them that the herbs will benefit their health. I have always admired the dogs kept by the gypsies of yesteryear, and believe them to be amongst the finest canine examples I have ever seen. These dogs will have had fresh herbs gathered from the hedgerows and herb-rich pasture added to their diet on a daily basis. The gypsies believed that it was the use of herbs that kept their dogs healthy. In order to have dogs to rival theirs, we too must make use of specific herbs.

The most important herb that you can give your terrier is garlic. This will strengthen the respiratory system, aid digestion and protect against internal parasites. There is a company that produces a herbal supplement called 'Keepers Mix'. This was originally devised by an old Dorset gamekeeper who wished to produce a herbal tonic that would have a vitalizing effect on many of the dog's major organs, with a particular leaning towards the stimulation and strengthening of the digestive system. It also contains celery seed, considered a time-honoured herbal treatment for muscle and joint pains, which will be of obvious benefit to the hard-working terrier, placing as it does enormous demands upon its body, and on its limbs in particular.

Minerals

Minerals have such a large array of function that they could fill an entire book in themselves, but it should suffice to say that all the cells in the body require minerals of some kind in order to carry out their specific tasks. One job that minerals do perform, and which is worth highlighting, is their vital involvement in the process of absorbing the circulating food nutrients. This shows us that although it is possible to provide your dog with a first-rate diet as regards protein and carbohydrates, if the mineral content of the diet is inadequate then the absorption of all that good food will be poor. With the working terrier, the efficient utilization of its food is essential if health and energy levels are to be maintained and, as a result, the mineral content of the diet has to be spot on.

It is also worth mentioning that too high a mineral intake can be as bad as a deficiency. For this reason, you should avoid adding mineral supplements to the diet of your dog, unless your vet instructs you to do so.

To my mind the best feeding regimes are those where the mineral and vitamin content occur in sufficient quantities within the raw ingredients. The easiest way to identify the essential minerals, and the foods in which they can be found, is by means of a simple chart.

Vitamins

Vitamins neither build cells nor produce energy, and yet they are essential for health because they have protective properties that guard the body from all manner of disease. Only a very small quantity of vitamins is needed, but if they are absent from the diet, the result will be ill health. Again, a chart is the best way to identify the vitamins, what they do, and in which foods they can be found.

Mineral Chart

Name of Mineral	Function of Mineral	Dietary Source
Calcium	Gives strength to the bones and teeth; also essential for muscle contraction, nerve function and blood clotting	Tinned fish; milk; white bread; bones
Phosphorus	Gives strength to the bones and teeth; essential for energy due to its ability to metabolize fats and carbohydrates	Milk; eggs; vegetables; cereals
Magnesium	Essential for chemical conversion of carbohydrates into energy; production of bone; movement of fluids within the body; neuro-muscular activity	Milk; seafood; whole grain; green vegetables; bones
Potassium	Body fluid regulator; also vital for all muscle activity especially the heart where it helps stimulate rhythmic beating	Meat; whole grains; tomatoes; peas; potatoes
Sodium	Transmission of electro-chemical impulses along the nerves; essential for fluid balance	Milk; bread; Weetabix; meat; fish
Chlorine	Produces hydrochloric acid in the stomach, which is an essential chemical for the breakdown of food	Milk; bread; Weetabix; meat; fish
Iron	Provides body with energy; transports oxygen in the red blood cells to all parts of the body; prevents anaemia	Egg yolk; liver; kidney; cereals; bread
Copper	Works alongside iron in the transport of oxygen within the red blood cells	Bread; cereals; meat
Zinc	Maintains healthy skin and helps to heal wounds	Wholemeal bread; cheese; egg; milk
Iodine	Helps regulate metabolism by keeping the thyroid gland healthy; boosts immune system	Fish; seaweed
Cobalt	Aids vitamin B12 with the production of new blood cells with bone marrow; prevents anaemia	Milk; eggs; liver

Fats

Fats are an important but not a large part of the dog's diet, with an estimated 5 per cent fat content being sufficient for most dogs. Fats are essential for the production of energy, the preservation of body heat and the normal function of the cells. They are also important as a vitamin reserve and an aid to digestion, and they contribute to a healthy coat.

The dog does not need large amounts of fat in its diet if it is being fed the correct amount of carbohydrates, because these can to some degree be turned into fat within the body of the animal. Carbohydrates will be the major source of energy for the dog, but an equivalent

Vitamin Chart

Vitamin	Function	Source
A	For vision in dim light; also health of the skin and the linings of the respiratory and digestive tracts; prevention of infection	Liver; egg yolk; milk; halibut; liver oils
B	Aids nerve function and production of new cells; helps with the extraction of energy from food	Wholemeal bread; wholegrain cereals; eggs; yeast
C	Promotes healthy skin	Orange juice; tomatoes
D	Vital for the dental and skeletal development of the young dog; also promotes assimilation of calcium and phosphorus in blood and bones; regulates metabolism	Fish oil; sunlight on the skin
E	There is some controversy over what this vitamin does for the dog; it may promote a glossy coat and aid fertility; definitely destroys dangerous chemicals known as free radicals	Most foods have vitamin E, and the richest source is vegetable oil
K	Clotting process of the blood known as coagulation	Vegetables; cereals

amount of fats provides two-and-a-half times as much energy. However, too many fats in the diet will lead to diarrhoea, and only active hunting dogs, such as rabbiting terriers, can tolerate slight increases in this part of the diet because they use up so much energy.

The body of the dog needs to build up some fat reserves, and it stores these underneath the skin, between the muscles and around the internal organs. These provide some insulation from the cold, and can afford a level of protection to the vital organs in the event of injury. One important fact about fat is that it takes a long time to digest when compared with proteins and carbohydrates, and it has a role to play in delaying the sensation of hunger. Therefore, a small portion of fat in each meal will make the meal more satisfying to the dog. I like to add more fat to my kennelled dogs' food during the winter months, as I feel the psychological effect of feeling full helps them to cope with the cold.

There is no doubt that dogs enjoy the fat part of the diet, and valuable sources of digestible fat can be found in milk, particularly goat's and sheep milk, eggs and meat. If extra fat is required in the ration, then butter, cream, suet and lard may prove useful. If you were to feed your dog on only the best, most expensive lean meat, he would look underfed, and the condition of his skin and coat would suffer due to the lack of fat.

Fibre

Fibre is the indigestible part of the food that aids the movement of food through the digestive system and helps in the correct formation of waste products. It is believed that fibre has a role to play in the maintenance of good health, but dogs do not require it in large quantities. Fibre is present in cereals, bread and vegetables.

Sugar

Most dog owners consider sugar to be harmful and avoid giving it to their dogs. However, there is no harm in giving very small amounts of sugar to the dog when it is needed, because it is a pure source of energy that can be instantly absorbed by the body.

As I mentioned earlier, sugar appears in the liver, muscles and blood of the animals on which wild carnivores feast. Consequently, a small amount of sugar is a naturally occurring foodstuff in the carnivore's diet and should not do the dog any harm. I provide my dogs with goat's milk on a daily basis, and this contains sugar in the form of lactose. Some people have reported a lactose intolerance in their dogs, but I have never had any problems whatsoever. Sugar is also useful when you want to supply your terrier with an instant burst of energy, and you can either opt for a natural source, such as honey, which dogs love, or use a powdered form of glucose.

Carbohydrates

Carbohydrates should constitute between 50 and 70 per cent of the working terrier's diet, and should provide the animal with the boundless energy needed for the sport of rabbiting. Even when the terrier is lying down quietly, energy is being used. When it stirs and exercises its muscles, the demand for energy increases, and this is met by carbohydrates that have been broken down in the terrier's digestive system, into glucose which, in turn, offers the terrier an immediate supply of energy. Carbohydrate-rich foods include pasta, rice, porridge oats and vegetables.

Raw Ingredients

Having discussed the essential elements of the diet for the rabbiting terrier, I shall now provide an example of the various ingredients that I regularly give to my dogs. It is worth bearing in mind that dogs can be just like people and tire of the same food at every mealtime, and I have found that dogs maintain much more interest in their food when an assortment of different ingredients is included. I thought that my grandfather was being a bit stingy when he insisted that my grandmother feed the dog with a small portion of the food that they themselves ate. However, not only did this provide his dog with a varied and interesting diet, it also contained, in far better measure than in most prepared dog foods, the essential components of the healthy diet.

There is a tendency today to underestimate the wisdom of old countrymen like my grandfather, as we become obsessed with research

Carbohydrates provide the terrier with the vital energy that it needs in its work.

findings as a basis for the way we treat our animals, and yet I have never seen healthier dogs and chickens than those kept by my grandfather. One of the most famous writers on dogs during the past one hundred years was A. Croxton-Smith OBE, and in his book *About Our Dogs* he states 'the house dog should have all the table leavings saved for him, and if these do not provide sufficient flesh, plenty of scraps may be had from the butcher for a few pence.'

The ingredients that I use are as follows: goats' milk; herbs (particularly seaweed and garlic); vegetables; household leftovers; mushy peas; canned pilchards or sardines in tomato sauce; wholemeal bread; pasta; wheat biscuits; and rabbit – I feed rabbit raw and on the bone, with the fur still on, to my dogs. If your dogs have been raised on cooked meat and you want this to continue, place the skinned rabbit cut in half into a large pan, cover with water, and cook slowly on the stove for approximately two hours. When it is cooked, remove all the meat from the bones. Cooked bones, unlike raw bones, should be discarded because the cooking process makes them brittle and very dangerous. The water that the rabbit was cooked in should be left to stand so that it can turn into a jelly, which dogs love.

How to Feed One Rabbit over a Seven-Day Period

Day 1
Liver; kidneys; heart and lungs

Day 2
1st front leg

Day 3
1st hind leg

Day 4
Half the body, cut behind the ribcage

Day 5
2nd front leg

Day 6
2nd hind leg

Day 7
2nd half of the body

Feeding the Terrier on the Day of a Hunt

Sample Menu

Early morning feed:
2fl oz of goat's milk with a level teaspoon of honey and glucose stirred in

Noon feed:
As above, with the addition of a quarter teaspoon of salt. The yolk of a boiled egg or a piece of cheese the size of a matchbox can also be given

Afternoon feed:
A good portion of rabbit, peas and plenty of water

Early evening feed:
As on a normal day

Once you have selected the *ingredients* you wish to feed your dog, you then need to start thinking about the *quantity* of food that you should give him. Over-feeding is a problem amongst modern owners, and a dog can eat as much as 50 per cent more food than it actually needs and, ironically, still be hungry. It is estimated that a dog weighing between 2.25 and 4.5kg (5 and 10lb) will need 250 to 600 calories a day. Another estimation is that the working terrier will require about 320g (11.5oz) of food per day. In order to make sense of these figures I have included a sample menu for the day of the rabbit hunt, and one for normal everyday use.

THE CARE OF THE ELDERLY WORKING TERRIER

The Old Terrier can Still Hunt

Some people think that when a rabbiting terrier slows down it should be retired and put out to pasture, but I believe the old terrier still has a lot to offer, and further, wants to

Everyday Menu

Time	Monday	Tuesday	Wednesday	Thursday	Friday	Saturday	Sunday
0800 8.00 am	goat's milk 2 fluid ounces	goat's milk 2 fluid ounces	goat's milk 2 fluid ounces	goat's milk 2 fluid ounces	goat's milk 2 fluid ounces	goat's milk 2 fluid ounces	goat's milk 2 fluid ounces
1200 Midday	half wheat biscuit with 2 fluid ounces goat's milk	half wheat biscuit with 2 fluid ounces goat's milk	half wheat biscuit with 2 fluid ounces goat's milk	half wheat biscuit with 2 fluid ounces goat's milk	half wheat biscuit with 2 fluid ounces goat's milk	half wheat biscuit with 2 fluid ounces goat's milk	half wheat biscuit with 2 fluid ounces goat's milk
1600 4.00 pm	1 slice bread offal from rabbit (liver, kidney, heart, lungs) 2 teaspoons peas 2 teaspoons orange juice herbs	1 portion of rabbit 1 slice of bread 2 teaspoons peas 2 teaspoons orange juice herbs	2 eggs scrambled or boiled 1 slice of bread 2 teaspoons peas 2 teaspoons orange juice herbs	100g fish 1 slice of bread 2 teaspoons peas 2 teaspoons orange juice herbs	2 eggs scrambled or boiled 1 slice of bread 2 teaspoons peas 2 teaspoons orange juice herbs	1 portion of rabbit 1 slice of bread 2 teaspoons peas 2 teaspoons orange juice herbs	1 portion of rabbit 1 slice of bread 2 teaspoons peas 2 teaspoons orange juice herbs
1900 7.00 pm	50g rice pudding	1 slice of bread 1 teaspoon raisins	1 slice of bread 1 teaspoon stewed apples	1 slice of bread 1 teaspoon raisins	1 slice of bread 1 teaspoon stewed apples	1 slice of bread 1 teaspoon raisins	50g rice pudding

continue its work for as long as possible. Imagine yourself, if you can, so enthusiastic about your work that every time it is mentioned your whole body quivers with excitement, and like a horse champing at the bit, you cannot wait to get started. Then one day, somebody decides that you should be retired because you are no longer as quick as you once were: so how would you feel? Somewhat peeved, to say the least, and your terrier is not immune from the same feelings.

There is plenty of stimulating work that an old terrier can do without being pushed beyond its limits. For example, I used to use my old terrier to pursue rats and even mice around the smallholding, which meant that she did not have to walk far to get a bit of sport. She may not have been 100 per cent successful, but the main point of these activities was to offer her some stimulation. Ratting, mousing or an hour's ferreting every now and then were all used to keep her mind challenged and body active.

In common with the North American Indians, and many other indigenous peoples of the world, I believe that animals are more than just flesh and blood. I believe that they have a spirit that is a part of their being and which gives them their individuality. If you have owned a number of dogs, you will know that no two are identical; each possesses its own distinct personality, which is as individual as a finger print. Bodies and minds may grow old but the spirit does not, and this is why working dogs never lose their desire to pursue their work, and why total retirement is just not an option. The elderly terrier, even when it is slowed down by the ailments that accompany old age, still needs to have its spirit roused, and this is achieved by allowing the dog to continue being useful.

I have seen the want of stimulation in old horses that have been parked in their stable or field as if they were a rusty old bike. They stand motionless, with a vacant expression and staring eyes that are bereft of both joy and excitement. Ageing may slow things down, but the old terrier has a fund of experience and wisdom to draw from, and should not be confined to years of boredom, or ignored because of the introduction of new blood in the form of a younger dog.

When working the elderly terrier, it is important that we do everything in our power to reduce the physical stresses that sport places upon the body of the dog. You can start by classifying your hunting sites as either easy, moderate, or demanding. The old terrier should only be worked on easy to moderate ground, which includes grassland and open woodland, leaving the more testing sites, such as areas covered in brambles, undulating terrain and rock ground, for younger terriers. By carefully selecting a hunting location, you will be able to keep the rabbiting terrier active during the advanced stage of its life.

The distance travelled to get to the hunting site is the next thing to be considered. During ferreting, it is not uncommon for me to travel many miles on foot as I journey from one warren to another in search of rabbits. When my terrier was young she covered the distance with ease, but as she advanced in years, all the walking became a bit too much for her to cope with. I resolved this by carrying her from one warren to the next in an old game bag, which meant that she could do all her work at the warrens and did not have to struggle with long distances. In this way she could continue to enjoy a full day of sport, without exhausting all her physical reserves.

You must also take into account the adverse effect that certain weather conditions exert on the body of the elderly terrier. Joint pains are exacerbated by cold and wet conditions, and the ability of the terrier to maintain body heat is also reduced. Avoiding foul weather, particularly if you live in the far north where I do, is impossible because all the serious sport takes place during the most inclement months. I overcame this problem

The elderly terrier's working day is prolonged by having a rest in this game bag that my brother carries from warren to warren.

by providing my terrier with protection from the weather in the form of a waterproof dog jacket. You may think that this is pampering a working dog, but it will enable the old terrier to work in comfort throughout the worst winter months – and from my own observations, it is always appreciated.

The terrier's mental appetite for work is apt to exceed its physical capabilities, and as a result the owner should keep a careful eye on his dog and stop it working itself to exhaustion. I remember my old terrier discovering a rat burrow, which she promptly began to dig out. After fifteen minutes a ton of earth and stones had been shifted. When I thought that she had done enough, and in order to prevent her suffering stiffness the next day, I called her off, much to her disgust. I then went to do some work in the stable. I returned an hour later to find my determined terrier back at the rat burrow in an enormous great crater that could have been made by a JCB. Enough was enough, and I picked her up by the scruff and shut her inside. If I had not intervened, she would probably still be digging!

A terrier of any age will be reluctant to stop working on the day of a rabbit hunt, and a firm command will be required. The trick is knowing when to finish so that the dog is stimulated without being too stiff and uncomfortable the next day. As long as my dogs can still walk, I will keep finding them some little job to do. My brother says that when my terrier is dead, I will still put her lead on and drag her round the field for some exercise. This is a bit of an exaggeration, but I do most fervently believe that an active, healthy mind and contented spirit can only be maintained if the dog's body is placed in challenging and stimulating environments. Old dogs that are kept still and quiet by the fire will soon lose their vitality.

Dietary Considerations for the Elderly Terrier

As a terrier makes its steady march into old age, its body will undergo a series of changes, foremost of which are a slowing of the digestive system, loss of muscle strength and slight reduction in the function of the liver and kidneys. It is therefore senseless to treat the elderly terrier in exactly the same way as you did when it was young, and a number of small alterations to the diet will be of immense benefit.

The first area that needs addressing in the diet is the consumption of protein. Many dog foods use protein in the form of meat or soya, to provide the dog with energy; this may be fine for the younger dog, but is not what is wanted at all for the elderly dog. For the dog to turn protein into energy, its liver has to change the composition of the protein and its kidneys have to deal with the waste products produced by the process. In the elderly terrier one of our main goals is to relieve the liver and kidneys from as much pressure as possible, and one of the key ways to do this is to reduce the protein intake. There is evidence to show that the removal of protein from the diet can improve the health of the liver and kidneys – with kidney failure being a major cause of death in elderly dogs, the more we can do to protect these organs, the better it is for the dog.

I would recommend that the meat portion of the diet be drastically reduced or maybe removed, depending upon the type of meat you feed your dog and the general state of the dog's health. There are numerous examples of dogs, including working Afghans and sheep dogs, which have been perfectly healthy despite the fact that they have never had a morsel of meat in their entire life. The problem today is the tendency of people to spend much more money on meat for their dogs and consequently end up with meats that are too high in protein values for the elderly dog. When feeding the elderly terrier, the intention is to provide an energy source that places the minimum of strain upon all the systems of the body, and this is accomplished by a low protein diet. The best protein and the most easily assimilated is that found in milk and eggs – the dog is able to absorb 95 per cent of the protein found in them. These egg and milk proteins are fed for bodily maintenance and not energy, and thereby place very little demand on the liver and kidneys; this, in turn lengthens the working life of the terrier.

Goat's milk is, in my opinion, the finest source of protein that you can give the elderly dog. The fat and protein particles in goat's milk are smaller than those found in cow's milk and are, as a result, more easily digested. Goat's milk has also been attributed with a number of health benefits amongst people who have allergies and those with a compromised immune system. For a long time goat's milk has been the milk of choice for orphaned animals and sick dogs. I implement the low protein diet by steadily reducing the meat intake and replacing it with goat's milk, most commonly made into a rice pudding, and boiled or scrambled eggs mixed with pasta.

A vital part of the elderly dog's diet is the herb, garlic. This is because garlic has an amazing ability to prevent high blood pressure, which is a common problem amongst the aged. During the course of experiments on rats, it was observed that garlic was able to lower blood pressure within twenty minutes of it being eaten. Garlic also has a role to play in the prevention of blood clots, and protects against respiratory and gastric infections. This is of obvious benefit because the elderly, just like the very young, are more susceptible to infections. Garlic also destroys worms, discourages ticks and fleas, and has properties that relieve the chronic pain associated with rheumatism. All told, it is the perfect herb to give strength and offer relief to the elderly working terrier, and it could not be easier to give. You can buy prepared garlic granules from any decent saddlery, or you could finely chop a clove of fresh garlic and add it to the terrier's feed.

THE RABBITING TERRIER AND KENNELS

The rabbiting terrier will live quite happily in the home and enjoy the opportunity to stick his nose into every aspect of family life. For many owners, having a bumptious little tyke constantly at their heels is an essential reason for keeping a terrier. However, this does not mean that owners who prefer to keep their terriers in kennels cannot make them

Kennelled terriers should have the stimulation of company and plenty of exercise.

perfectly happy and contented. The kennelled dog, just like the house dog, requires adequate amounts of exercise and the companionship of its owner. I would advocate that at least two terriers be kept together in a kennel, because to my mind it is not fair to keep a dog with the terrier's companionable disposition in a kennel all by itself.

The discussion of kennels and kennel management could fill a book in itself, and the following serves only as an introduction to the rudimentary principles regarding keeping a dog in a kennel.

It is well documented that when human beings live in poor housing they are far more likely to suffer mental and physical ill health and, in some cases, have a shortened life expectancy. The same applies to dogs and their kennels, with badly designed and poorly managed kennels having a devastating effect on the health, happiness and longevity of the working terrier.

To understand what the essential requirements of a good kennel are, we shall spend a moment looking at that familiar country figure, the badger. By so doing we are following the example of the gypsies who acquired a careful and prolonged observation of nature. Badger setts, as you obviously know, are dug into the ground, and there is good reason for this. Not only does the earth provide shelter from wind and rain, it also soaks up the solar rays of the sun, day in and day out, whatever the weather and whatever the season. If you have watched the endless round of house programmes, you will no doubt have seen that this latent heat absorbed by the earth can be extracted to heat houses. By making his home below the ground, the badger, in his infinite wisdom, has been utilizing this solar heat for thousands of years, the result being a sett that is cosy and warm, even in the depths of the British winter.

Sadly, many dog kennels do not provide anywhere near this amount of comfort, and many a working dog would gladly swap its miserable kennel for a simple burrow in the ground. The temperature in a wooden kennel during the winter can easily drop below freezing, as is evidenced by the thick layer of ice on the dog's water bowl the morning after a deep freeze. Some of the thicker-coated terrier breeds may be able to tolerate these temperatures, but nature shows that exposure to such conditions is not the best option. Badgers, foxes and rabbits make their homes deep beneath the earth where such temperatures do not exist. I read about a man who built his house into the earth and it was so warm that he never needed any heating, irrespective of the harshness of the weather outside.

So how can we keep the terrier's kennel just as warm and stop it freezing? The answer is simply to install some form of heating, and this is by no means a new concept. In books that I own dating from the late eighteen-hundreds, there is information to be found on the heating of hunt kennels by means of coal. The best option today is infra-red heating, because the infra-red rays warm the dog rather than heating up the air. If you use an infra-red lamp, make sure to suspend it at least a metre (4ft) above the standing terrier in order to eliminate the risk of the lamp burning the dog, which it will, if placed near the animal. This type of heating is easy to install and cheap to run, and in return it will literally add years to the working life of the rabbiting terrier, as well as vastly improving its comfort during the winter months.

Badger setts are not only heated, they are also well ventilated, by which I mean that the foul air is expelled by the badgers as their breath leaves the sett and is replaced with fresh air from the outside. You may be wondering how I know this. If you have ever been around a badger sett you will doubtless have seen what looks like smoke rise from it. This is not the badger busily sending messages, but hot expired air leaving the sett and condensing as it meets the cold air outside. The reason why the foul air rises so accommodatingly to the surface is simply because it is warm and, as any student of basic science will tell you, gases when heated become lighter and rise. The hot air balloon is a perfect example of this principle.

When the hot foul air exits the sett, it leaves a vacuum – basically an empty space, like an empty car-parking spot. Just as another motorist will instantly fill this space, so the vacuum left by the foul air is instantly filled by cooler, fresh air from outside the sett, which it rich in oxygen. If we have the good sense to learn from nature, we must seek to apply this principle of ventilation to our dog kennels, so that our dogs may, like the badger, have constant access to fresh, clean, richly oxygenated air.

Ventilation is so important because it maintains a clean environment by removing the dirty air that has been breathed out, and replacing it with fresh, unpolluted air. When a dog breathes, coughs or sneezes, droplets of saliva and secretions from its nose are sprayed into the atmosphere. If the dog is harbouring any infection, this spray will contain pathogenic organisms, which are the vehicles that transport disease from one animal to another. They can travel 6 to 10m (6 to 10yd) and may infect any animal that happens to be in range, which is why it is so important to clean the air by means of ventilation.

In a poorly ventilated environment, pathogenic organisms can also fall to the ground where they dry and mingle with the dust, to be inhaled or ingested at a later date. Sweeping the floor in such a kennel actually compromises the health of the terrier by stirring up the organisms that have fallen to the floor, which makes cleaning the kennel an absolute waste of time. It should also be noted that the warm, moist atmosphere in an unventilated kennel provides the perfect conditions for the rapid multiplication of any germs present, and can weaken the respiratory defences of the dog. Therefore an unventilated kennel is guilty of actually attacking the health of the dog, firstly because it is a virulent breeding ground for infectious organisms and secondly because it actively reduces the physiological ability of the dog to fight infection.

When installing ventilation, you will need an exit point through which foul air can escape, and this is thankfully quite easy to achieve due to the fact that this air rapidly rises to the highest point in the kennel. Therefore, by situating an opening of some nature at the highest point in the kennel, foul air will readily exit of its own accord.

Fresh air will rush in to fill the vacuum left by foul air, and if we do not provide an entry point for this incoming air near ground level, it will rush in through the roof vent and in the process create an uncomfortable and unhealthy down-draught from which the poor dogs will not be able to escape. Having no ventilation in a kennel is detrimental to their health, but even worse than this is ventilation that has been badly incorporated and creates draughts. Your dog would do better to lie outside than to live in such a kennel. Thus in addition to the exit vent in the roof, we must also have an entry point for fresh air.

We can now turn our attention away from the design of the kennel to its actual management, and the first practice I wish to challenge is that of washing down kennels. Some years ago I used to re-home ferrets from a very large animal rescue facility that had kennels covering an acre or more, and which must have cost many thousands of pounds to construct; this facility employed a large staff, including veterinary nurses. Each morning the kennel staff would hose down the kennels and runs, spraying gallons of water everywhere. During the summer, the kennels would eventually dry, but once the weather began to turn a bit cooler, they never dried out thoroughly. The dogs themselves always seemed to have damp paws and never looked happy, and despite the presence of veterinary staff, kennel cough was rife.

I am quite convinced that washing down kennels has no beneficial effect whatsoever on its occupants and is, in fact, injurious to the health of the dog, and particularly those that are elderly and require a dry environment if their muscles and joints are to function to their full potential. It has long been recognized, as a result of countless studies into the effects of poor housing, that damp conditions encourage the spread of respiratory infections. Moreover, anyone who has dug out their ferret from a rabbit warren will tell you that wild animals, such as the rabbit, do not live in damp conditions, and their burrows are, in fact, completely dry – and this is because all wild animals know that dampness causes discomfort, pain and disease. The daily washing down of kennels may, on the surface, seem like a good idea, but it is, in reality, a destructive practice. Water and kennels, like oil and water, just do not mix. A horse makes a far larger mess than any dog, and yet I do not know of a single groom who washes out their stable on a daily basis.

When you do not rely on the use of water, you can clean the kennel in the following way. To start with, the kennels should have a deep layer of wood shavings over the entire floor, excepting of course, the run. The initial laying of such a bed may require several bales of wood shavings, which cost in the region of £5 each, but afterwards you will only need a bucketful each day to set things in order. With a deep bed of shavings like this, cleaning is a simple matter of scooping up any droppings and urine-soaked areas, and adding fresh shavings to the places from which the fouled material was removed. In this way the kennel is kept fresh, hygienic and sweet-smelling because shavings have a very pleasant smell and, most importantly of all, the kennel is kept dry.

In the run, remove any droppings, and then take a bucket of piping hot water, to which you should add a few drops of tea tree oil. Tea tree is a natural disinfectant, and is used in hospitals in Australia. Place a stiff, long-handled brush into the bucket so that you get just the bristles damp, and then shake these over the bucket until they are almost dry. You can then use the brush to sweep the run, returning periodically to the bucket in order to keep the brush fresh. In this way, you will end up with a clean, hygienic run that is totally dry moments after the sweeping is completed.

The thick layer of shavings on the kennel floor will also prevent the terrier coming into contact with a bare concrete floor. Concrete floors are uncomfortable and cold and quickly become damp, and householders are understandably quick to cover them over with carpets. The shavings serve the same purpose as carpets and provide cushioning for the terrier when it is resting, and as a result, helps prevent stiffness of the limbs and muscles. The pressure on bony prominences when they are in contact with a hard surface may lead to the eruption of small sores on the skin. Consequently, a deep carpet of shavings is a preventative strategy and well worth the investment by anyone who wants to get the best out of his working terriers.

With the whole of the kennel floor covered with shavings, we are in a position to make some additions that will provide the dog with a warm and comfortable place to sleep. You could place a generous amount of straw in one corner, but for the terrier I favour putting the straw in what I refer to as a 'tunnel'. This is basically a long box with several small openings through which the terrier can squeeze. The idea of the tunnel is to provide the terrier with an artificial den in which it can rest. Some people provide some kind of raised platform for their terriers to sleep on, but these do not give the terrier as cosy an environment as the tunnel.

As bedding, nothing, in my opinion, is better than straw, although as a knowledgeable dog keeper pointed out to me, shredded paper shares the same properties as straw and is an excellent choice for those terriers that have an allergic reaction to straw. Straw is one of the warmest beddings that I know of; in times gone by, hot food would be taken from the kitchens of the great houses to the shooters and beaters in boxes lined with clean straw, and the food always arrived steaming hot because straw has such a marvellous insulating effect. It can work just the same with a terrier, when a foot-deep bed of straw is provided, which the little creature can burrow into. Once the terrier is well snuggled in, the insulating properties of the straw ensure that its body heat is not dissipated.

Straw not only keeps the terrier warm, it also draws moisture away from its body, which means that a terrier that has been working in the rain and has then bedded itself down in straw, will be both warmed and dried by this material. A large number of man-made beds are now available on the market, but nothing beats good old-fashioned straw.

Kennelled terriers can therefore live happy and contented lives, but this will depend on the owner giving them adequate amounts of exercise and companionship. It could be argued that kennelled dogs actually require more care and attention than their house-dwelling kin, and it should never be viewed as a system where the dogs are left to their own devices except for feeding times and the occasional working excursion.

SIGNS OF HEALTH AND ILLNESS IN THE RABBITING TERRIER

You will be surprised by how much information the terrier can provide you with regarding its health; all you need do is use your eyes and hands. Terriers are amongst the longest lived and healthiest dogs and are the easiest of creatures to keep fit. Even so, every terrier man should monitor the health of his dog so that he can identify any sickness at its earliest stage, and have it treated promptly. Sometimes dogs have to endure long operations or even be euthanased because they were not taken to the vet early enough. The value of a weekly health check for the dog is therefore obvious.

Monitoring the health of your terrier is therefore a simple matter of observing the following:

- respiratory rate;
- pulse rate;
- food/fluid intake;
- fluid output;
- elimination;
- condition of mouth;
- general condition.

We shall now look at each of these briefly, in turn.

Respiratory Rate

Respiration is the process whereby air is breathed in to supply oxygen to the rest of the body. It is a process that comprises inspiration, when the external muscles between the ribs and the diaphragm are contracted enabling air to be drawn into the lungs, and expiration when air is breathed out as these muscles relax. The respiratory rate is the number of times that this occurs within one minute and, although this varies greatly among the different breeds, an average is estimated as eighteen to twenty-five breaths per minute. Observing the respiratory rate of your dog when it is healthy will provide you with its normal rate, which you can then use as a benchmark for comparisons.

You can identify the respiratory rate of your terrier by watching the movement of its chest when it is at rest. This movement is quite obvious to the eye, and you will see it expand or broaden and then return to its normal size; this counts as one breath. The respiratory rate will increase with exertion, but any other sustained alteration to the breathing pattern of your dog should be noted and a vet consulted in order to identify the underlying cause. Difficulty in breathing is very uncomfortable and is a source of anxiety to the animal experiencing it. Also take note of persistent coughs.

Pulse Rate

The pulse rate is the rhythmic expansion of an artery, which may be palpated or felt when the artery is near the surface of the skin and passes over a bone. In humans it is the radial artery of the wrist that is most

An example of a healthy terrier showing good muscle tone and an alert expression.

commonly felt, but the best way to take the pulse of a dog is to press your fingers against the inner fleshy side of the hind leg well above the knee. At first you may have to feel around a bit, but having once found it, you will be able to locate it quickly on future occasions. The average pulse rate for dogs is 70 to 100 beats per minute, with the smaller dogs, such as terriers, having a faster rate.

The pulse rate is an indicator of cardiac or heart function because the expansion of the artery corresponds to the contraction of the left ventricle of the heart. With the terrier, a sustained increase to 120 beats and above should be reported to the vet because it could be indicative of an underlying disease or poor heart function. When taking the pulse it should feel strong and conform to a regular pattern. As with breathing, the pulse will increase with exercise or if the dog is placed in a stressful situation.

Food/Fluid Intake

Food and water are essential for life, and an alteration in their intake by the dog is a sure sign that something is amiss. Every dog owner is aware of his animal's pattern of feeding, and will be able to recognize changes in this pattern quite quickly. If the dog fails to resume his normal dietary regime within a day or two, the vet should be consulted because there are so many conditions where loss of appetite is a prominent feature. Do not be concerned if your dog eats grass and then vomits a small amount of mucous bile, because this is quite natural and affords relief to the stomach.

The owner should also keep a daily watch of how much water his dog is drinking. If your dog suddenly begins to drink copious amounts of water and seems to have a thirst that cannot be satisfied, you should take it to the vet to determine the reason for this. For example, polydipsia (the technical term for abnormal thirst) is a classic symptom of diabetes.

Fluid Output

Just as you keep an eye on what goes into your terrier, you should keep a check on what is coming out. In normal health the dog should pass urine regularly and without discomfort. If this function becomes uncomfortable, or the dog seems to be urinating constantly, you must find out the reason for this change. Polyuria, or the abnormally large output of urine, is attributed to an excessive intake of fluid or to disease, most commonly diabetes.

Elimination

This refers to the bowel action of the dog, and the owner should ensure that his dog is opening his bowels regularly without any discomfort and that the waste that is passed is correctly formed. Loose bowels and diarrhoea should be carefully monitored because although they could signify something simple, such as indigestible food or worms, they could be a sign of something much more serious, such as enteritis or poisoning.

Condition of the Mouth

Check that the colour of the gums is pink, and that there are no loose teeth. Apply gentle pressure to the gums, and observe for any signs of pain. Also keep an eye on the build-up of tartar on the teeth, and use your sense of smell to determine the state of your dog's breath. If this is offensive, you should give the dog a thorough mouth-wash, and if after this the smell is just as strong and constant, you should go to the vet because it could be indicative of an underlying condition.

General Condition

You should observe the brightness of the eyes, the moisture of the nose, the lustre of the coat and the movement of the limbs. Use your hands to inspect every part of your dog's body for lumps, cuts or hair loss. Pay attention to the mood of your dog and its responses to you. They may sound like insignificant events, but when they are combined with other changes in your dog they will help to build a picture of what is wrong with a sick dog, and a good vet will appreciate all the information that you can give him. Furthermore, such observations may make you aware of minor ailments that you might have overlooked. Many of the minor ailments fall within the category of home treatment, if the owner is confident, and the dog will be grateful for such judicious care.

FIRST AID TREATMENT OF THE RABBITING TERRIER

It is the custom with most dog books to give a long list of diseases that can affect the dog, and which will require a vet to correctly diagnose and treat. I have chosen instead to concentrate on first aid treatment that can be

The healthy terrier must be free-moving and typically display a spring in its step.

administered by the terrier man to his dog when it suffers the inevitable bumps and cuts that accompany a sporting life. The following is just a basic introduction to canine first aid, and the terrier man would be well advised to study the subject in more depth because the nature of the terrier's work exposes it to the risk of injury. I have concentrated on the more common injuries.

First Aid Kit

For the Field

- 1 roll of surgical tape
- 1 small jar of honey – an antibacterial with no side effects, as well as promoting wound healing
- 2 K bandages
- 1 crepe bandage
- 2 non-adherent absorbent dressings known as Release
- 2 face wipes
- Tweezers
- Blunt-nosed scissors

For the Car

- 5ml syringe
- Isotonic fluid
- Warm or space blanket
- Flask of warm water
- Cotton wool
- Bowl

Bruising

This is an injury of the deeper layer of the skin in which the capillaries of the blood vessels are ruptured, causing an outpouring of blood within the skin, which in turn gives the characteristic discoloration that we associate with bruises. A really severe blow can cause the bruise to penetrate the underlying muscle, and those parts of the body that are not covered by a layer of muscle, such as below the knee and the hock, are susceptible to bruising of the bone, for example if the dog collides with a solid object, which sometimes happens if the terrier jumps over a fallen tree or wall. A bruised bone will thicken as it repairs and the dog may display some stiffness, but this will quickly wear off in the younger dog. However, the bones of the elderly dog are not as resilient, and an untreated wound may leave the dog with rheumatic pains.

The bruise will more than likely be treated when you return from your day's rabbiting, but if you should wish to treat the bruise in the field, the application of something cold is the correct course of action and will reduce the swelling. Alternatively, you could carry Hilton Herbs hot and cold treatment which can be used for deep heat/ice pack effect, and consists of oils of peppermint, birch, juniper and wintergreen, and should soothe any bruising.

At home the treatment is as follows: rub some arnica gel into the bruised area and then apply a warm compress to relieve the pain. A hot water bottle wrapped in a towel will suffice for this. If the bruise is accompanied by a break in the skin, first cleanse the wound with water, apply a small amount of honey, which both disinfects and promotes healing of the wound, and then proceed as before, making sure not to rub the arnica into the open sore because it will sting. Limited exercise should be taken until the stiffness caused by the bruising has worn off.

Torn Ear

Due to the nature of rabbiting and the construction of the ear of the dog, there is always the possibility that the ear could get torn; the dreaded barbed wire is the most persistent offender with regard to this injury.

If this does happen, the ear will bleed profusely, making it look far worse than it actually is. Simply take a dressing from your field kit and apply it to the tear, which should stop bleeding. If it does not, or if the tear is more than half an inch long, you will need to bandage the terrier's ear to its head, and then go to the vet. To bandage the ear, first fold the ear upwards over the head with the dressing covering the tear, and then wrap a bandage over the ear and around the neck several times.

Scratched Eye

Scratches of the eye should be bathed with water that has been boiled and left to cool until it is lukewarm. You can then gently bathe the eye with cotton wool, though avoid wiping the surface of the eye. It is much better to flush the eyeball with water, and you can do this with a pipette. It is helpful to have an assistant to keep the dog still. After bathing, wipe the whole area dry with a fresh piece of cotton wool. A drop of castor oil can be placed on the eye, and this will have a soothing effect as well as forming a protective layer over the eye.

Broken Nail

Breaking a nail is a fairly common occurrence in sporting dogs, and it is important always to be aware that this injury can be both very sensitive and painful. The nail may be snapped in half, completely torn off, or left dangling from the root. If the nail is cleanly broken off, drip a small amount of honey over it as a disinfectant, and then proceed to carefully bandage a small dressing on the site of the injury. If the nail is split and needs removing, place some cotton wool round the injury and use some surgical tape to hold the wool in place. This should suffice in the short term, and enable you to get to a vet who can administer a local anaesthetic to deal with the pain while the injury is treated.

Thorns in the Pad

Dogs that pursue rabbits through cover on a regular basis can easily pick up a thorn in their pad; this is not a great cause for alarm, even though the terrier will be lame until the thorn is removed. Anyone who has watched a terrier nibble away at its feet will know that the dog will try and pull the thorn by itself.

You must attend to any offending thorns immediately, and you should be able to do this with a pair of tweezers. A magnifying glass can also prove useful if you cannot see the thorn. You should grasp the head of the thorn with your tweezers, and with one clean movement, pull it out. Do not move it from side to side because it may break and the point will still be embedded in the pad of the dog and is then much more difficult to remove. A hot compress can be applied to draw the thorn to the surface so that it can be removed. If this does not work, either apply a bran poultice or visit the vet. A bran poultice left on for an hour should succeed where perhaps a compress failed, and bring the thorn to the surface. The dog will show immediate relief when a thorn is removed, and should be able to walk normally.

Cut Pad

It is quite common for terriers to cut their pads on, for example, broken glass, which appears far more in the countryside than you would expect. As soon as you become aware of this type of injury, roll your terrier over, apply direct pressure to the wound and then secure an absorbent dressing in place with a bandage. When you return home you can remove the dressing. Take care that you do not tear the dressing off the wound because this will start it bleeding again and delay its healing. If a dressing is stuck it is much better practice to soak it in warm water: this will eventually release it without hindering the healing process. Warm water is used because it keeps the wound at the best temperature to promote healing. If the wound has stopped bleeding, smear it with a little honey and redress it, but if the wound will not stop bleeding, apply a clean dressing and take the dog to the vet.

Wounds

A wound is an opening in the skin, and there are a number of different types of wound including incised, laceration and puncture wounds.

An Incised Wound

This is a clean cut such as you would receive from a knife. The terrier may get cut on glass, metal, barbed wire or briers, and a wound of this nature cuts through the flesh and blood vessels, resulting in a considerable amount of

Here we see a location where rabbits have built a warren under dumped material. Note the barbed wire and sharp bits of metal that could put the terrier at risk and cause injury.

bleeding. If such a wound is more than 2cm (1in) in length, it may require sutures. Obviously, if an incised wound strikes a major blood vessel, it is an emergency situation and you should get to the vet as quickly as possible.

The first objective in the treatment of an incised wound is to halt the bleeding, and this is done initially by the application of direct pressure, following which a dressing should be bandaged over the wound. As you secure the dressing, you should try to draw the wound edges together.

The bleeding of a wound may be either arterial or veinous, which simply means that the blood comes from an artery and vein respectively. Arterial blood is bright red in appearance, and will seem to spurt out regularly, corresponding to the beat of the heart. This is because the blood is in a pressurized system, with the heart being the source of the pressure. Arterial bleeding is more likely to occur in the limbs of the dog than any other part of its body. To treat this bleeding, two bandages will be required: the first should be tied firmly above the wound like a piece of string, which will stem the flow of blood from the wound; the second bandage is used to secure a dressing over the wound itself. Then take the dog to the vet at quickly as possible.

If you have not reached the vet within half an hour, take the bandages off and examine the wound for one minute to see if the bleeding has stopped. If the wound is still bleeding, reapply the bandages as previously described, only move the first bandage to a position slightly higher or lower than it was before. Keep an eye on your dog for signs of shock, and every five to ten minutes offer a drink of fluid that is one pint of water with a tablespoon of glucose and a teaspoon of salt dissolved in it.

Blood from a vein is dark red in colour and will gush constantly. A dressing held in place by a bandage should be immediately applied to the wound; if the bleeding does not stop, tie a bandage below the wound, but only leave this in place for ten minutes at a time. Keep the terrier warm and administer isotonic fluid for shock. As is always the case, if the bleeding will not stop, get to the vet without delay.

Lacerated Wounds

This type of wound is the most common to be seen in sporting terriers, when the skin is torn open. This often happens because the terrier squeezes through or under obstacles where the skin is dragged along the ground. Tight spaces and the friction between the skin and the surface of the obstacle result in small tears in the flesh. These wounds do not bleed much because the blood vessels are stretched rather than ruptured. Clean the wound with cotton wool soaked in warm water. When you clean the wound, use a fresh bit of cotton wool each time you wipe over the surface so as to avoid contaminating the wound. When the wound is clean, put some honey in the centre of your dressing and then apply it directly to the wound. Secure the dressing with a bandage or surgical tape, depending where the wound is on the body of the dog.

Basic Rules regarding Wound Management

- Always use warm water when washing a wound: this enables it to stay at the best temperature to promote healing.
- If a dressing is in situ, avoid changing it unless it is necessary; you may think you are doing the best thing, but in fact you will cause small amounts of damage and delay the healing process.
- When a dressing is stuck to a wound, always soak it so that you can remove it without tearing the new tissue that is forming under it.
- Change a dressing if there is fluid, known as exudate, showing through it. Some wounds will seep, but this is nothing to worry about, and you should try to use a dressing that is made of a material that does not stick. When changing a dressing take note of the size and depth of the wound, and the presence of any offensive smell.
- If the wound continues to smell, consult the vet. You can get dressings that are impregnated with charcoal and designed for use on wounds that stink.
- Do not underestimate the value of honey in wound management: it may sound like a strange thing to use, but research by doctors and nurses has proved its value in the treatment of wounds.
- Not all wounds require dressing, and you should allow this sort of wound to take its natural course and hopefully heal itself, just as *you* would if you had a small cut in your hand.

Puncture or Stab Wounds

These wounds occur when a sharp object, such as a nail, pierces the skin and penetrates the underlying tissue. Some puncture wounds bleed severely and can continue to do so despite the application of pressure and dressings – in which case the dog must be taken to the vet without hesitation. For less dramatic puncture wounds first-aid treatment will suffice. You should begin by trimming any hair that hangs over the wound, and then, with tweezers, remove any visible fragments of wood, glass and suchlike. The wound can then be bathed with cotton wool and warm water, following which a bran poultice should be applied for one hour; when this is removed, a small amount of honey should be massaged into the wound area. This process of wound management can be repeated for three days. The purpose of the poultice is to fight bacterial growth by drawing any infection out of the wound.

Eye Washing

It is important to wash out a dog's eyes on the day of the hunt because terriers that go to ground often return to the surface with their faces covered in dirt. Brush this off with your hand, and gently wipe away any dirt from around his eyes using one of the individual face wipes you can buy these days. When you return home or to the car, you can then wash the terrier's eyes thoroughly with a flask of warm water poured into a bowl. Use cotton wool to wipe over the eyes, but if there is any dirt in the eye you will need to use a small syringe to very gently flush the eye with water. Have an assistant hold the eyelids open while you do this, and keep the dog calm and still. You can administer one drop of castor oil to soothe the eye when it is clean and your dog is relaxed.

Further Reading

Lucas, Captain Sir Jocelyn, MC
The Sealyham Terrier; Hunt and Working Terriers (Tideline)
Jocelyn Lucas was a veteran of World War I and was awarded the Military Cross. He went on to become a Member of Parliament. He was an active sportsman during the first half of the last century, and founded a well known kennel that bred Sealyhams. He was an expert on all aspects of terrier work, and worked a large pack of Sealyhams.

Plummer, Brian
In Pursuit of Coney (Boydell)
The Complete Jack Russell Terrier (Boydell)
The Sporting Terrier (Boydell)
The late Brian Plummer is probably one of the most well known workers of terriers and lurchers in recent times, and wrote constantly on these subjects for *The Countryman's Weekly*. He was a biologist and teacher who gave extra-mural lectures on wild life in several universities. He once survived in the Midlands for eighteen months relying on hunting. His books are full of information and are always interesting to read.

Russell, Dan
Working Terriers (Tideline 1980)
Dan Russell is an ex-Fleet Street journalist who first became interested in terriers when he was eight years old, and was given his first terriers when he was ten. He worked terriers for the Enfield Chase Foxhounds and helped the Old Berkeley Hunt. He went on to become terrier man for the Exmoor Foxhounds. He wrote regularly for the *Shooting Times* and was a strong supporter of the Jack Russell and Border terriers.

Smithson, Bob
Rabbiting (The Crowood Press 1988, out of print)
Bob Smithson worked as a gamekeeper, and then as a teacher of environmental studies including botany and ornithology. He is a sportsman with a proven record, and has an extensive knowledge of all methods of rabbiting. He is of the old school, sportsmen who make all their nets instead of buying them. He retired in 1983. He recalls catching 10,600 rabbits in one year when he was nineteen years of age, and he did this with a team of ten ferrets and two lurchers.

Index